# ARTFUL

# SCIENCE

ENLIGHTENMENT ENTERTAINMENT AND THE ECLIPSE OF VISUAL EDUCATION

*Barbara Maria Stafford*

*The MIT Press*  *Cambridge, Massachusetts*  *London, England*

This book was set in Garamond by DEKR Corporation and was printed and bound in the United States of America.

Library of Congress Cataloging-in-Publication Data

Stafford, Barbara Maria, 1941–
    Artful science : enlightenment entertainment and the eclipse of visual education /
Barbara Maria Stafford.
        p. cm.
    Includes bibliographical references and index.
    ISBN 0-262-19342-6
    1. Optical illusions—History—18th century. 2. Scientific
recreations—History—18th century. 3. Mathematical recreations—History—18th
century. 4. Visual perception. 5. Art, Modern—17th–18th centuries.
6. Scientific literature—History—18th century. 7. Science—Study and
teaching—History—18th century. I. Title
QP495.S73 1994
306.4′5—dc20                                                                                        93-20984
                                                                                                              CIP

Illustration credits are given in the List of Illustrations, beginning on page viii.

*For Fred*

INTRODUCTION

This is a short book about a large topic. It looks at the "mathematical recreations" and "philosophical entertainments" that sprang up during the long eighteenth century. These exciting ways of doing science by stimulating the eyes were devised to amuse and improve a broad spectrum of European society from the baroque to the romantic era. Informative as well as entertaining, illustrated popular books, optical cabinets, marvelous machines, astonishing experiments, and provocative museum displays contributed to the swelling stream of public pedagogy, adult education, and the recuperation of childhood that crested during the Enlightenment.

Recent scientific discoveries that early stimulation is crucial for brain development have led to the reassessment of the role played by sensory experience in knowledge formation. The brain uses the outside world to shape itself and to hone such crucial powers as vision, reasoning, and language. Not hard wiring but continual interaction with the external environment is now thought to produce even the most abstract kinds of cognition. This notion that the harder you use your mind, in youth or in age, the more in shape it will be was anticipated by early modern manuals and practices for intellectual exercise. These graphic entertainments, relying on seeing, hearing, touching, and smelling, laid the foundation for future cognitive development. Stocking the imagination with countless images that could be disassembled and reconnected into innovative patterns, engaging teaching tools made consciousness synonymous with the continual processing of the physical world.

In seeking to historicize entertainment, I want to focus on the *visual* component of the burgeoning eighteenth-century leisure industry. While much fine scholarship has been done on the textual and literary aspects of recreation flourishing in the resorts, spas, and urban centers of Europe, surprisingly little attention has been paid to subtle sensory forms of knowing in the birth of popular education as amusement. Researchers into the rise of private life and the inception of consumer culture have recounted how, approximately after 1700, economic and political factors combined to produce a widening and prosperous middle class. More and more people had the means to travel at home and abroad, go to the theater, browse fairs, attend lectures, build laboratories, gape at cabinets of curiosities, buy fashionable gadgets for themselves and the latest toys for their children. While zealously setting about "improving" their homes and gardens, they also yearned to improve themselves less taxingly through the joys of looking and reading.

Searching analyses of literacy abound, and the important findings of economic, social, literary, and science historians are acknowledged throughout the following pages. Yet the mind-shaping powers of ocular, tactile, kinesthetic, and auditory skills remain scarcely articulated in the tale of Western civilization's turn to the cultivation of the interior. The fast-growing field of book history continues to focus primarily on reading habits and textual interpretation. When images are introduced into such studies they tend to serve as illustrations of changes in the meaning of writing, not as vibrant shapers of knowledge. Uncovering this lost epistemological dimension of the informed and performative gaze, and with it the complex interface of early modern nature and artifice revealed in moments of enlightening recreation, seems all the more important in our computer era. Now old intellectual traditions based on crayons, loose-leaf paper, and paste are also being replaced by playful high-tech tools and visually appealing programs: Windows, mouse, PC Paintbrush, Excel, and PowerPoint. The rise of electronic media casts print culture, as well as the histories of art and science on which these disciplines are grounded, in sharp relief.

Telling the story of the beginnings of mass literacy by focusing on the eclipse of visual aptitude is to introduce fresh materials, new conceptual models, and forward-looking technology into traditional accounts. In a book also about the explosion of experimental science, I experiment by focusing on representative moments in the singular transit from an oral-visual culture, to a literate one, to the uncanny recurrence of pictures ever more lifelike within vivid multimedia performances.

Not only is the use of graphics on the rise in home computers, but they have invaded the classroom. In 1989, Chris Whittle launched Channel One, a controversial television news service that gave public and private schools a free twelve-minute current events program each day. This gift was accompanied by free satellite dishes, free cable wiring, free videotape units, and free television sets. One promise was exacted in return. Students were all but required to watch both the news show and the two minutes of commercials embedded within it. Looking toward the millennium, Whittle dreams of creating one thousand technology-driven schools funded and stocked by a handful of corporations.

There are enormous implications to Whittle's desire to program a young audience through an electronic curriculum, but I wish to address only his ostensible motivation: the fact that public education is in a shambles and that students and educators alike perceive schools as ineffectual and boring. Instead of attempting futilely to retrieve nineteenth-century ideals and standards of textual literacy, it might be helpful to focus on the other side of the historical story. Rather than looking at newspapers, pamphlets, magazines, and books for the evidence of eighteenth-century readership, I look at a wide range of graphic materials that were the ancestors of today's home- and place-based software and interactive technology. I argue that high-order thinking was taught in the construction of visual patterns and that optical technology often boosted the learning process of difficult abstractions.

Whittle's notion of generating ideas that inform and entertain in several media is not new. Nor would the debates over whether such schemes were merely money-making operations or revolutionary visions have struck an eighteenth-century audience as novel. Indeed, there was a protracted struggle by their French, German, Italian, and English inventors to distinguish sensory diversions for leisure hours from idle pastimes and vicious sports. Consequently, instructive games must be considered within the broader artistic, social, and intellectual contexts of an ostensive ornamental science. Indicative of this universal "taste for experiment" was the increasing staging of spectacular trials, the exhibition of curious and showy natural history specimens, and the use of images to solve abstract problems concretely.

This book originated in my thinking about the deep connections of the eighteenth century to late modernity. Will the ash end of modernism be followed by the Reenlightenment? Struggling with that question, and that of the changing role of visual studies in the era of global electronic conduits and the optical distribution of information, resulted, first, in a state-of-the-field essay for the 1988 *Art Bulletin*. In addition, the present work amplifies my book-length investigation into the metaphorology of the body, and of phenomenal experience in general, during the eighteenth century. These concerns remain all the more pressing today in light of the proliferation of "distance-learning" courses via videotape, satellite, cable, telephone, and computer. Thousands of busy working people are currently taking college and postgraduate classes for credit using VCRs and PCs. The lure of "telecourses," mind extension services, and "digital professors" makes it all the more important to argue for the need to *talk* together about images as well. A fine line separates technology as a necessary supplement to education from telecommunications' opportunism and media packaging.

Juggling, *tours de force,* legerdemain ("the visible invisible"), and conjuring constituted the province of the charlatan. I identified this "visual quack" with the profiteering sophist in *Body Criticism.* Licit and edifying manual operations, however, were not always easy to distinguish from illicit and duping fairground routines intended to mystify largely illiterate spectators. We are plunged, then, into the vernacular oral-visual world of the *ancien régime.* Its credulous denizens were being alternately harangued and enticed into self-improvement. The *philosophes'* antireligious and aniconic polemics were directed against "Oriental despotism"—including that of the Roman Catholic church—supposedly holding its subjects in "Asiatic" thrall by the seductive lures of voice and sight. Charismatic hucksters equipped with divine gimmicks remain an enduring social phenomenon. The honey-tongued preacher of Sinclair Lewis's *Elmer Gantry* and the fast-talking Reverend Jonas Nightengale of the recent movie *Leap of Faith* continue to fascinate worldwide audiences by their faking, paranormal claims, rigged "miracles," and faith healing.

The eighteenth century offers compelling evidence for just how complicated the verification of authentic experience was and still is. The intervention of instruments could produce optical illusions that were greeted with delight or skepticism. Empiricism unsettlingly resembled pseudoempiricism. Playful tricks with light, stealthy sleights-of-hand, macabre body tricks contorting science demonstrators, eerie and obsessive automata, all depended upon a sensuous technology. Mediating instruments, both useful and recreating, were the staple goods of the professional engineer, the ingenious inventor, as well as the defrauding mechanic. The blurring of distinctions between the organic and the artificial lives on in the media-saturated, postindustrial cyberpunk. Fans of William Gibson's *Neuromancer* (1984)—whose themes include prosthetic limbs and implanted circuitry—might reflect on Vaucanson's attempts to alter apparatus into flesh. Computerism, which is rapidly ousting all forms of modernism including postmodernism, is remarkably rococo in its doubleness. Then and now, ambivalence greets the paradigm-shattering role of technology as producer of the joys and terrors of communication.

In uncovering this ambiguous realm of artful science—lying somewhere between entertainment and information, pleasure and learning—I wish to move past the nihilism of postmodernism with its text-based epistemology. No one who has watched the computer graphics and interactive techniques revolution can doubt that we are returning to an oral-visual culture. Animation, virtual reality, fiber-optic video, laser disks, computer modeling, even e-mail, are part of a new vision and visionary art-science. What is lacking today, however, is a concomitant high-level visual education to accompany the advances in visualization. Broadly trained experts in all facets of the history, theory, and practice of graphic design must join with imaging scientists to teach the abiding and changing tactics for the creation and perception of simulations, including those appearing on screen.

I think we can better equip ourselves for the revolutionary universe of marginless Ethernets by exposing the sophistication of an earlier age's wrestling with the problems of manipulation and distortion, phantom and illusion. Visual education, I argue, arose in the early modern period. Significantly, it developed on the boundaries between art and technology, game and experiment, image and speech. The exchange of information was simultaneously creative and playful. We need, therefore, to get beyond the artificial dichotomy presently entrenched in our society between higher cognitive function and the supposedly merely physical manufacture of "pretty pictures." In the integrated (not just interdisciplinary) research of the future, the traditional fields studying the development and techniques of representation will have to merge with the ongoing inquiry into visualization. In light of the present electronic upheaval, the historical understanding of images must form part of a continuum looking at the production, function, and meaning of every kind of design.

But to return to the pattern of this book. Chapter 1 takes up the oral-visual culture of the late baroque and examines how it became tarnished during the Enlightenment. The evolution of mathematical recreations is traced within this shifting milieu. Instructive games developed uneasily from the witty, arcane, and private conceits of an aristocratic court culture to the rational entertainments of the middle class who increasingly demanded edifying public amusements. The problem of filling the interval between work and rest was especially acute in a society bedeviled by hoaxes and scams of all sorts, whether artistic, scientific, religious, or political.

Chapter 2 reveals that many diversions were devised to teach the unwary how to guard against fraud. In a revision of the customary dualism, I suggest there was a fundamental continuity between the Enlightenment moral preoccupation with sincerity and the romantic ethical quest for authenticity. The eighteenth-century battle against charlatanism, delusive machinery, and "technological speech" developed into the early nineteenth-century attack on virtuosity. What is marvelous, extravagant, or extraordinary is very often the result of astonishing manual skill, disturbingly capable of creating both genuine and ungenuine effects.

In an epoch in which specialization and professionalization were just getting under way, the public and frequently heroic performance of experiments could appear embarrassingly like a magic show. Chapter 3 examines those ambiguous attempts at visual persuasion that were part sublime hocus pocus and part egalitarian "science of everyman." Trials were staged by a new and equivocal caste of *amateurs en nouveautés,* demonstrators, and self-testers. Trying out experiments on oneself was an extension of the Enlightenment's activist program for progress.

By devising a different account of the eighteenth-century interactions between art and science, I identify the development of an experimental painting rivaling the natural philosopher's laboratory games. Artful scientists, in turn, downloaded cosmic energy into their wired and gadgetized bodies with ever greater pictorial realism. An incarnational pedagogics taught beholders through living examples. Portraits and genre paintings—from Greuze to David, Kauffmann to Maron, Wright of Derby to Gainsborough—tangibly demonstrated an encyclopedic gamut of skills and behavior. In entrepreneurial quarters, then, representational interest had shifted by midcentury from the product's essence to a viewer-oriented process. Figures, rather than passively wearing their attributes, empirically instructed consumers how various actions were physically done.

The optically based rhetoric of being versus becoming can be situated within the larger debate surrounding exhibitionism. Experimentation conspicuously reconfigured both matter and self. These promethean uses of technology also had an impact on systems of ordering. Were natural objects and human artifacts to be arranged ostentatiously according to their flamboyant materials, or reticently disciplined by linguistic schemes? Chapter 4 looks at the dilemma of display at one of its primary sites, the natural history cabinet and museum. An antinomy existed between spectacle, or gawking at heaped-up goods, and observation, or the reasoned apprehension of phenomena. Just as graphic recreations were supplanted by serious textbooks communicating increasingly specialized information, and experimental philosophy came to occupy a lower cognitive rung than the quantified physical sciences, so the installation of objects grew to be dominated by the linear trajectory of the catalogue.

The difficulties attendant upon collecting and classifying allow us to examine, from another vantage, the tensions inherent in the very concept of pleasurable information. Artists, experimentalists, amassers of machines, artworks, and specimens struggled to define just the right relationship between entertainment and instruction, the balance between an aesthetics of watching and the rigors of intellection. In a society gravitating from looking to reading and fragmented into diverse audiences—specialist, amateur, *curieux,* gaper—filling the vacancy of leisure demanded increasingly heterogeneous visual arrangements. Museums, like the mathematical recreations to which they should be compared, confronted the problem of the expanding "interval" by the accumulation and layout of rarities.

At present, information is in transit, crossing over from "hedonistic" oral-visual modes to "serious" textual methods and back again. Manufacturers of the hottest new digital devices aspire to turn the planet into a virtual garden of sensual delights. Confronted by all this vaporware and techno-hyperbole, media opponents have reproduced, without realizing it, the Enlightenment critique of merely beautiful appearances. Like the early moderns they, too, deride the charlatanism of a bewitching optical technology and a conning manual skill without compelling content performed to trick the masses. My conclusion, on the contrary, claims it is not fated that the interval allotted in the past to recreation must now become a seamless continuum of besotting entertainment.

When, on June 3, 1769, the sun was obscured by the passage of the planet Venus across its luminous face, expedition scientists using instruments calculated what they could not bear to look upon. Classification and quantification were the Enlightenment's rational methods for controlling a mythic and carnal hypervisibility. Symbolically it seems that, today, it is Venus—concupiscent goddess of illusory appearances and promiscuous pleasures—who casts her shadow over every male cognitive domain. Although this transit from a verbal to a visual culture is profoundly unsettling to postmodern iconoclasts, it is a reminder that there is no learning without desire, no education without enjoyment.

In writing this monograph I have incurred many debts. I wish especially to thank the Alexander von Humboldt Foundation for awarding me a Senior Fellowship in 1989. This continuing grant helped me to write not one but two books! I am also deeply indebted to the wonderful staff and facilities at the Herzog August Bibliothek at Wolfenbüttel. Dr. Sabine Solf made my stay there not only productive but memorable.

In the United States, I was fortunate to be invited to spend a quarter in the spring of 1991 at the University of California Humanities Research Institute at Irvine. This project was born in the lively meetings of our group on Aesthetic Illusion, guided by Frederick Burwick. I thank our energetic leader and the collective intelligence and wit of Reginald A. Foakes, Marion Hobson, Deanne S. Howe, Donald G. MacKay, Walter Pape, K. Ludwig Pfeiffer, Elinor S. Shaffer, Wayne Slawson, Thomas Vogler, and David Warren. Mark Rose, Director of the Humanities Institute, and his staff made it a joy to be in residence.

Likewise an intellectual delight, my brief stint as a Visiting Getty Scholar, in April 1991, was immensely profitable. Thank you to Martha Reed for her friendly expertise in locating and obtaining rare volumes, and to Kurt Forster, Tom Reese, and Herbert Hymans for generously incorporating me into the life and mind of the Center. An invitation by James Winn to be a guest at the University of Michigan Humanities Institute in the spring of 1993 proved propitious. The warm and friendly atmosphere conducive to scholarship was created not only by the director but by Mary Price, Eliza Woodford, Betsy Nisbet, and Linnea Perlman. In addition, I am grateful to George Rousseau and David Summers for their helpful comments on the manuscript.

Anselmo Carini, Associate Curator of Prints and Drawings, Art Institute of Chicago, remains both resource and guiding angel. Suzanne Ghez, Director of the Renaissance Society, was an inspiration in her dedication to contemporary art and issues. At the University of Chicago, too, Robert N. Beck, Director of the Center for Imaging Science, has been deeply supportive. His vision of imaging science is global enough to include the fine arts. Alice D. Schreyer, Curator of Special Collections, Joseph Regenstein Library, was very accommodating. Philip Gossett, Dean of the Humanities Division, kindly helped to defray photographic costs. Thank you also to art history graduate students who kept me on my toes: Anna Arnar,

Katherine Haskins, Elizabeth Liebman, and Marina Belozerskaya. Linda Nguyen, my undergraduate assistant, was not only inexpressibly helpful but a benevolent presence easing literary loneliness. As always, my muses Hildette Rubenstein and Ann Seeler were an inspiration. I thrived on the friendship of Claire and Stan Sherman, Susan Zimmerman, and Leeds Barroll.

The biggest debt is the last. My husband Fred urged me, for a change, to write a short book! I would not have thought of the problem of what it means to be experimental—whether in art or in science—without his living example and enduring support.

All translations unless otherwise indicated are my own. Book illustrations are cited in the captions first by author and second by engraver.

ORPHEE,
ou les Effets du Discours.
Frontisp. de la Gramm. Univ.

L'Harmonie en naissant produisit ces miracles.
Boileau Art Poétiq.

1

Antoine Court de Gebelin

*Orpheus, or the Effects of Speech*

1774

from *Le Monde primitif*

## "Oriental Despotism"

Physical objects, including images, have fallen into disrepute. Social historians ask whether looking at them can possibly constitute a major source of knowledge.[1] Anthropologists deplore the supposed tyranny of vision, based on the optical appropriation of thinglike data, and choose to tap instead into an invisible realm by studying narrative.[2] Students of material culture, while realizing that commodities are a type of information, identify them with European imperialism and the attempt to dominate nature.[3] In the process of trying to get at why people insatiably desire goods and engage in "an apparently endless pursuit of wants,"[4] the conquest of "serious" texts by "hedonistic" images is often the presumed culprit. An endless stream of digitized diversions has become synonymous with mental vacancy, the mind's release from hard reality and human responsibility. Literate activity, according to many media critics, is in danger of being overrun by a succession of consumer-driven amusements communicated sensationally. Thus television's principal contribution to educational philosophy has been summed up in the negative judgment that teaching and entertainment are inseparable.[5]

The metaphors of graphic despotism used by contemporary authors are startlingly reminiscent of those employed in an earlier time. Mid-eighteenth-century northern Europe (especially France, Germany, the Low Countries, and Great Britain) was in the throes of changing from an oral, visual, and aristocratic culture to a market-centered, democratic, print culture. Supposedly ignorant listeners and gullible onlookers had to be molded into silent and solitary readers.[6] While the transformation in late modernism runs in the opposite direction from texts to images, the pangs of this culture-making metamorphosis were registered in similar analogies during early modernism's trajectory from images to texts.

Bruno Latour has coined the term "black boxing" to describe those assumptions in science that are taken as givens, that are presumed no longer to require discussion and thus have become invisible.[7] In this book, I want to open one leaky black box and examine its heterogeneous contents. What were the social influences and cultural biases that contributed to the demotion in status not only of an entire "southern" oral-visual mode of learning[8] (fig. 1) but of an attendant train of not always obviously associated "baroque" trades, genres, and styles? These included optical technology, ingenious experiments, exhibitions of curiosities, theatrical

2

Rembrandt van Rijn

*Joseph and Potiphar's Wife*

1655

performances, *trompe-l'oeil,* and digital dexterity. A broad gamut of extraordinary, marvelous, and crafty things were condemned in the wake of the perceptual skepticism cast by Hume and others. Modern Pyrrhonism rendered suspect not only miracles but all eyewitnessing. As in Rembrandt's prescient representation of the false ocular testimony offered by Potiphar's wife against an innocent Joseph (fig. 2), the misleading judgment of the eye came to be impugned in the eighteenth century. Corrosive doubt made visual evidence synonymous with legerdemain wrongly persuading the beholder there was proof of causal connection when, in fact, there was only probable conjunction.[9]

Today, we need to go backward in order to move forward. We must unearth a past material world that had once occupied the center of a communications network but was then steadily pushed to the periphery. This imagistic and casuistic empirical universe was crisscrossed by extra-linguistic messages, interactive speech acts, gestured conversations, and vivid pantomime. Spatial and kinesthetic intelligence were not yet radically divorced from rational-linguistic competence and logical-mathematical aptitude. Acquiring adroitness in the manipulation of symbols was a form of intellectual sport accrued by paying attention, by observing individuals at practice.[10] Objects, like works of art, might be brought together for the purpose of showing them to others. Thus they held the potential for education by inviting mutual participation in an enjoyable experience. As semiophores, or carriers of meaning, things could also bear witness to a positive and instrumental materialism; not as the passive drugs feeding our habit of consumption, as culture critics have defined products, but as cherished possessions. As delicate and personal tools, they stimulated that optimal psychic and somatic flow defined in the phenomenological psychology of Mihaly Csikszentmihalyi (fig. 3). Domestic items, instruments, images, specimens, toys—in fact, anything designated within the environment—could become significant by patterning existence and imparting order to disorderly minds.[11]

This interactive and fluid communications system was predicated on that momentary act of conjuring known as conversation. The British romantic writer Thomas De Quincey praised velocity in the movement of thought cultivated specifically within such eighteenth-century milieus as the salon, coffee house, and club (fig. 4). As Reynolds's group portrait attests, it is not necessary to assume that only profitable labor allows people to create strong and complex personalities. Conversation, as the acme of leisure

3

Jean-Baptiste-Siméon Chardin

*The Draftsman*

1737

4

Sir Joshua Reynolds

*A Conversation (George Selwyn, George J. Williams,*

*and the Honorable Richard Edgecumbe)*

c. 1759

activity, draws objects into the orbit of the evolving self. Visual performance is a kind of juggling or "moving amongst moving things," in the words of De Quincey. Fragments and snatches of speech coalesce into ideas that palpably grow and develop before the eyes.[12] Mobile thought is material, elastic, splintering into fresh forms and startling angles bodied forth in the discordant yet harmonious poses of Reynolds's energetic speakers.

The dynamics of colloquial discussion, I suggest, capture the range and kinds of abilities valued in an oral-visual culture. Frank colloquy means rehearsing one's intellectual perplexities among friends and colleagues. In filmic fashion, the disclosures and revelations of conversation demonstrate that the best means of learning are by teaching. Again De Quincey's remarks prove apt: "The readiest method of illuminating obscure conceptions, or maturing such as are crude, lies in an earnest effort to make them apprehensible by others." Unlike solitary reading, such oral-visual drama is practical reason in action but also the instantiation of Kantian disinterested delight. Conceptions are visibly generated, one can "play with them," as with tangible objects, "watch and pursue them through a maze of inversions, evolutions, and harlequin changes."[13]

Acrobatic repartee, however, teeters dangerously between being a skilled fine art and a mechanical "mountebank display." Gymnastic speech, reveling in dazzling elocution, like painted images flaunting their showy execution, was identified with vulgarity. The culture of politeness had no tolerance for ostentation. Witty talk—like the flashy tricks of the funambulist, the gawdy nostrums of the quack, or the colossal idols of Eastern tyrants—was ungentlemanly in its material conspicuousness.

Already in seventeenth-century religious polemics, the northern and Protestant drive to establish abstract reading and writing was meant to unseat "Romish traditions" relying upon gulling words and duping icons.[14] Throughout the eighteenth century, Catholicism remained synonymous with the popular culture of the common person, recalcitrantly immune to the edifying powers of print and depressingly prone to superstition and relic worship.[15] Significantly, Rembrandt depicted his Calvinist, Remonstrant, and Mennonite sitters (fig. 5) explicating and pondering the divine message sent by an invisible God. His material absence was symbolized by the chiasmal structure of the painting depicting the well-to-do shipping magnate and gifted lay minister Anslo preaching to his intently

5

Rembrandt van Rijn

*The Mennonite Preacher Anslo and His Wife*

1641

listening wife.[16] The Dutch master's dual emphasis on the need for interpretation and concentration highlights the visible gap separating human
readers from the Holy Scriptures and the intangible divinity concealed
behind them.

For Reformers, as scholastics of the Word, the book became a new object
of worship bringing in its train a cadre of close interpreters.[17] Conversely,
baroque painters from Caravaggio to Sir James Thornhill were castigated
for their pompous, alien, and Italianate abandon.[18] To Whig and Tory
alike, the papist emphasis on office, ritual, and the passions of the confessional threatened the freedom of individual understanding. Worse,
according to Dissenters and Protestants of all stripes, Roman Catholic
artists represented apocryphal miracles that seemed no better than pagan

6

Caravaggio follower

Healing of the Blind Tobit

c. 1615–1620

magic in disguise. To iconoclastic observers, Tobit's improbable cure from blindness, caused by the dropping excrement from a swallow's nest and involving the descent of the archangel Raphael and the attack of "a fish," seemed an untrustworthy adventure story (fig. 6).[19] The gall was less remedy than carnal object. This crude fetish corresponded to the animality of credulous believers capable of equating the miraculous intervention of the deity with the flashy showmanship of a carnival charlatan.[20]

The image of the eighteenth century as the secular Age of Reason has served to obscure the magnitude of the collective anxiety gnawing at the Republic of Letters concerning the collapse of an original monotheism into polytheism. This sensual "Orientalism" presumably corrupted a primitive "unitarianism" and was fueled by the juggling knacks of a mystery-mongering priesthood.[21] The kinds of religious and ethnic stereo-typing occurring in the Enlightenment polemic against opulent satraps, memorably recorded in Delacroix's *Sardanapalus* (Paris, Musée du Louvre, 1828), need to be analyzed. From Hume to Boulanger and Coleridge to Renan, vitriol was thrown upon Moses "the Egyptian priest," Mohammed "the charlatan," and even Christ "the impostor."[22] These prestidigitating necromancers released the mind from rational thought by their distorting enchantments. I believe one cannot overemphasize the impact on all areas of artistic, scientific, and social thought of the concentrated Enlighten-ment attack on "Asiatic" theurgy. This artful science of specious demon-strations supposedly preyed on the stupidity of enslaved races fed daily doses of delusion. Further, the relentless critique of manufactured splen-dors and ocular falsifications is the single most important, yet overlooked, condition behind the demotion of images and speech.[23] Having been so thoroughly identified with the cruel technology of visual domination, they remain to this day erased as *positive* forces on the historical scene.

Nicolas-Antoine Boulanger's *Despotisme oriental* (1761) painted an unfor-gettable picture of the Near East as a collection of barbarous and tragic lands filled with people in a humiliating and deplorable condition of servitude. This volume did much to disseminate a knowledge of the Orient as it had been orientalized, refracted through the positional supe-riority of the European.[24] One hears this tacit dialectic between East and West in the French author's contrast of effeminate submission to hegemonic control as against the virile rule of right reason. In exotic countries, "man, destitute of will, adores his tyrant, kisses his chains." Boulanger's profession as an engineer and inspector for the Corps des

Ponts et Chaussées made him keenly aware of the useless suffering inflicted by the forced road labor of the *corvée*. Analogous to the contemporary situation in France, the absolute power wielded by ancient "Asiatic monarchs" was shored up by religion. Political despotism thus derived directly from a *theological* error. In remote ages, humanity took as its model of government the universe ruled by a supreme being. This faulty visual imitation "plunged all nations into idolatry and thralldom."[25]

I have examined in *Symbol and Myth* Boulanger's elaborate anthropological theory of a universal flood whose survivors established the cults, rituals, cultural institutions, and priestcraft in evidence among the early Hebrews, Chinese, Hindus, Persians, Phoenicians, Egyptians, Greeks, and Romans.[26] What is significant for our present purpose is that he interpreted all ancient patriarchal theocracies as commemorating primeval revolutions in nature. The expectation that the end of the world was imminent led to stargazing. Astrologers read the sky as if it were a legible book written in "arabesques" or Hebrew characters (fig. 7). These cataclysmic upheavals left the human mind disposed ever after "to be the dupe, the sport, and victim, of all those fanatics and impostors, endowed with cunning enough to conciliate to them the attention of mortals, ever actuated with a vague hope and undetermined expectation."[27]

Boulanger's *Antiquité dévoilée* (1766) expanded on the strategies operating within a theocracy whereby society not only worshipped the supreme being as God but came to believe that he was their direct and designated king. When the divine monarch was conceived to be like a man, politics became subordinate to religion and they mutually contaminated one another. The materialization of the deity relentlessly progressed. The God-King was given a house and it became a temple in which an effigy was placed. This image attracted the gaze and desire of the masses and thus became an idol. A table was set before the God-King so that he might eat and it was transformed into an altar. Attendants were required to immolate animals and to perform sacrifices. Theocracy was thus "a realm of priests," an "illusory government." Engulfing true divinity, it converted the God of monotheism into a cruel and evil tyrant forcing mankind into abject submission. Promiscuous false deities bred the bastard "demigods" of polytheism, who in turn arrogated divine rights unto themselves.[28] Reminiscent of Montesquieu's Usbek in the *Lettres persanes* (1720), these pseudodeities operated within a celestial seraglio.[29] The despotic system was thus also an oppressive sexual economy, subjugating the weak to the desire of the masters.

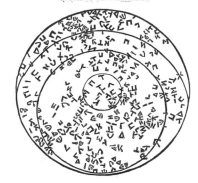

7

Jacques Gaffarel

*Configuration of Stars, or Celestial Characters*

1637

from *Curiositez Inouyes*

In the *Christianisme dévoilé* (1756/1766?), Boulanger affirmed his belief that despotism was established in all unenlightened portions of the globe and continued to flourish. Nowhere could its afterlife be better witnessed, he argued, than in the persecution, intolerance, and zealotry perpetrated by "Popish Christianity." The network of privileges, monopolies, prerogatives, immunities traditionally claimed for the Holy See had rendered its abuses notorious by the eighteenth century. Orientalizing the Catholic church, this friend of Diderot and d'Holbach discerned Eastern theurgy lurking behind the "puerile ceremonies" of transubstantiation. Continuing the *Encyclopédie*'s project of debunking, he mocked a God "coerced by the magic power of a few words, accompanied by rites, to obey the voice of His priests, or those who knew the secret of how to make Him act on their orders." Boulanger remarked that, even in his day, this charade was staged by Christian conjurors who persuaded their disciples they could command the divinity by means of spoken formulas to appear miraculously in the sacraments.[30]

Since the time of Luther and Calvin, "illusion," "false-packing," "cozenage," "knavery," and "deceit" were associated with the enchantments promulgated by Catholicism's "jugglers."[31] But after the Counter-Reformation, it was the Jesuits—politically controversial since their founding in the sixteenth century—who were specifically denounced by rigorist Protestants and reform-minded Augustinian Jansenists alike. Under widespread attack by the *philosophes* since the 1750s, they were finally expelled from France when, in 1764, Louis XV completed the dissolution of the order within his realm. Subsequently, high hopes were entertained for pedagogical changes, especially in secondary education, throughout Catholic Europe. The Jesuit system of collegiate theater had been roundly criticized by Rousseau, Helvétius, and d'Alembert as indulgent spectacle at the antipodes of proper training. Dressing up vice in showy garb and ostentatious rhetoric only served to feed the sinful inclinations of an adolescent audience enticed through pyrotechnics and bombast directed at eye and ear (fig. 8).[32]

To be sure, as Christiane Klapisch-Zuber has shown, the borderline between devotional and play activities has always been faint. Medieval marionettes resembled the figurines used to enliven preachers' sermons. In the Florentine quattrocento, *bambini* or effigies of male babies formed part of a devotional pedagogy. They were also incorporated into a domestic sacred theater in which infant dolls were manipulated by servants, women,

8

Francesco Solimena (?)

*Education of the Virgin*

18th century

François Hemsterhuis

*Destruction of Optical Optimum,*

*or Contrariety in Contour*

1792

from *Lettre sur la sculpture*

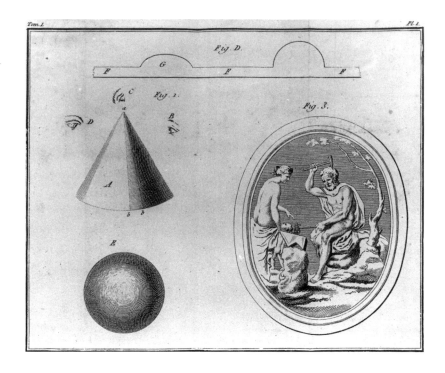

and children in games, rituals, and a type of interactive fantasizing.[33] Significantly, these emblematic puppets were intended to satisfy the needs of the less cultivated minds in the Renaissance household. Instructing the illiterate was accomplished through powerful visual impressions.

For eighteenth-century warriors against universal credulity, however, such diminutive sculpture was no different from the huge mounds of stone erected by "Oriental" patriarchs. The Dutch analyst of Greek scholia[34] and pro-Hellenic art theorist François Hemsterhuis, in his *Lettre sur la sculpture* (1789), contrasted crude colossi raised by ancient Near Eastern civilizations with the beautifully proportioned statues of the Greeks and Romans (fig. 9). As a student of confused and corrupt texts, it is not surprising that Hemsterhuis applied standards of historical criticism concerning authentic written productions to visual conglomerates.

He argued that within the small republics of the Hellenes every individual counted, and so a genuine personification reigned. Conversely, among the Egyptians, the theocratic ruler was everything and the mass of humanity merely an assemblage adding up to nothing. This wholly material atmosphere of political oppression was incarnated in the fabrication of marvel-

10

François Hemsterhuis

*Complex and Simple Vases*

1792

from *Lettre sur la sculpture*

ous effects and the manufacture of irrational monsters. Compounding animal heads and superimposing them on human torsos represented the height of contamination. Hybridism expressed an ignorant population's "esprit de symbole et de merveilleux."[35] More generally, for this Winckelmann-inspired Dutch Platonist, lingering over the physical properties of any object became a superstitious act. Accumulating minutiae, irrespective of the culture that made them, bred disgust in rational observers.

Hemsterhuis's insufficiently known theory of aesthetic perception proposed an immaterialist program for visual education. As was true for eighteenth-century hermeneuticists studying the commentaries of grammarians or the interpretations of the Bible, the reliability of a version rose when there were fewer interpolations. Artistic merit was also predicated on an "optical optimum" calculated according to the swiftness of the eye's travel over complete forms. Hemsterhuis's contrast of two visual "variants," an "Asiatically" ornate and volumetric vase with a Greek amphora whose decoration was controlled and separated into legible registers (fig. 10), must be seen in relation to the Enlightenment controversy over pure and impure editions. Significantly, he connected the complicated "Eastern" designs to Gothic, that is, Catholic, art defined as popular and

*Mala noche.*

11

Francisco de Goya

*Bad Night*

1799

appealing to low instincts. The critic made an additional comparison of the vase's monstrous confusion to a child's inept drawing of a horse. Typically, young and inexperienced draftsmen stopped to record every anatomical detail without seizing the sense of the whole. Most damningly, however, such "contrariety in contour" was equated with a "fantastical material voluptuousness."

In all three analogies, the eye was forcibly slowed or halted by a repugnant heterogeneous clutter. Like the toiling reader of messy glossaries, the perceiving soul yearned to behold a blotless page, an effortlessly flowing and substanceless line. In this sense, the corporeal minimalism and clarity of classical Greek art resembled the impalpable intellect immune to the corruptibility of matter.[36]

For enlightened reformers well into the romantic period, however, the Catholic Church was synonymous less with incoherent baroque sculpture and congested minor arts than with idiotic public processions, incantatory spectacles, *tableaux vivants,* and mesmerizing floats.[37] Goya's scornful critique of the foolish exhibitionism engaged in by a superstitious populace, enslaved by its credulity, shaped both his tapestry designs and print series (fig. 11).[38] *Los Caprichos* (1799) cinematically unfurled ironic scenes of delusion and charlatanism. Such visual diversions formed the secular antithesis to religious contemplation. Counter-Reformation strategies had urged the faithful to conjure up scenes from the life and passion of Christ during meditational exercises (fig. 12). Still reproduced in eighteenth-century emblem books for consumption in Catholic countries, such visions were interpreted by scholars of the occult as the acme of the modern thaumaturge's dubious and artful science.[39] Apparitions, sleights-of-hand, ventriloquy, perfumes, magical invocations—unveiled or mumbled within a mysteriously dim sanctuary—constituted everywhere and at all times the stratagems and machinery of "the empire of priests."[40]

The Belgian balloonist, physicist, and international prestidigitator Etienne-Gaspard Robert, in his *Mémoires récréatifs* (1831), unmasked the persistence of false experimentalism or "art des illusions." The counterfeiting of supernatural events in subterranean darkness stretched from the Egyptian mysteries, to the oracles of the Greeks and Romans, down to the divinations of the Druids. Significantly Robert (later known as Robertson) claimed to have rediscovered this ancient art of phantasmagoria in 1787. The cunning techniques for evoking nocturnal shades through

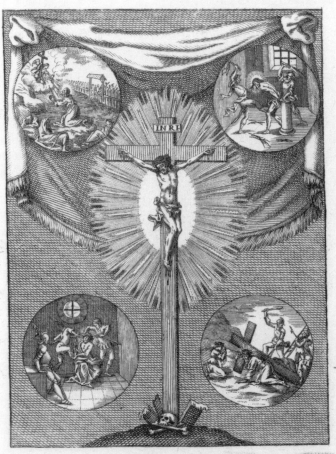

Rex æternus sæculorum,
tanquam latro, famulorum
   jussu comprehenditur;
flagris lacer, atque spinis
coronatus, in divinis
   membris crucifigitur.
Jener höchste HErr und König
wird den Dienern unterthänig,
   wie ein Mörder eingefangt,
wird gegeißelt und gekrönet,
mit dem größten Spott verhöhnet,
   endlich an das Kreuz gehangt.

Si te Christus non amasset,
putas tanta tolerasset,
   aut fudisset sanguinem?
hunc amorem meditare,
fædam tuam contemplare
   & ingratitudinem.
Hätt' sein Lieb sich nicht ergossen,
wär nicht soviel Blut geflossen,
   weder litt' er solchen Schmerz;
diese Liebe oft bedenke,
und dich schamroth tief versenke
   in dein undankbares Herz.

Illam lavâ salutari
in amoris Jesu mari
   exundante fluctibus,
qui de plagis salientes
mundant omnes pœnitentes
   a culparum sordibus.
Wasche dich in jenen Brünnen,
die aus JEsu Wunden rinnen,
   und dir bringen alles Heil,
siehe, wie sie stromweiß fließen,
abzuwaschen, welche büßen
   alle ihre Sünden-Beul.

S

12

Anonymous

_____

*Visualization of Christ's Passion*

1779

from *Ichnographia Emblematica Triplicis*

**OPINION.**

13

Jean-Baptiste Boudard

*Opinion*

1766

from *Iconologie*

machinery, however, had to be revamped for an illuminated age. He set out to transform duping ghosts and specters into pedagogical weapons in the arsenal of the Enlightenment. A true son of the *ancien régime,* Robertson maintained that the phantoms generated during his fashionable Parisian seances were not the lying appearances of old. Rather, he saw these "experimental trials," involving the concealed manipulation of mirrors, as akin to the initiatory rites of Freemasonry. Darkness, pain, and fear preceded illumination. Group epiphanies occurred under the myriad lamps installed in his theater of simulation housed in the former Convent des Capucines. Metaphorically, sudden brilliance symbolized the regeneration of mankind finally ushered into a philosophical Elysian Fields.

Until his death in 1837, Robertson crisscrossed Europe, taking his magic lantern to Russia, Germany, England, Spain (where Goya witnessed a performance), and Italy. These instructive shows, he declared, were to uncover the artifice of priests, both ancient and modern. This pernicious caste preyed upon "the grossly credulous," particularly in the centuries before the invention of the printing press. For, without it, even if someone had discovered their chicanery, there were no effective means for communicating that liberating information.[41]

The alleged "Orientalism" of the Roman Catholic church collided with more than the free-thinking and libertine proponents of a literate Republic of Letters, hoping to be rid at last of the tyranny of priests. Puritanism and an ascendant Protestant ethic of inwardness and reflection mounted an attack on sensual perception. The so-called "Second Reformation" of the late seventeenth century, initiated by Pietism, fostered new ways of reading. Instead of the Catholic view of books as dangerous sources of vain opinion (fig. 13), not to be consulted without ecclesiastical guidance, Lutherans and Calvinists encouraged private reading. The Bible, novels, and even *à la mode* forgeries like Macpherson's *Fingal* might be perused alone in a quiet place (fig. 14).[42]

Recent scholarship has revealed the magnitude of Pietism's impact on northern and central Europe, including the Catholic territories.[43] Unlike the imagistic psychology and epistemology of Counter-Reformation Catholicism (fig. 15), Pietist pedagogy celebrated an unimageable and invisible God. Remote from any taint of physicality, the Absolute was symbolized as the intangible play between proximity and distance in the mountain landscapes of Caspar David Friedrich (fig. 16). These austere

14

Anton von Maron

*Portrait of Archibald Menzies*

1763

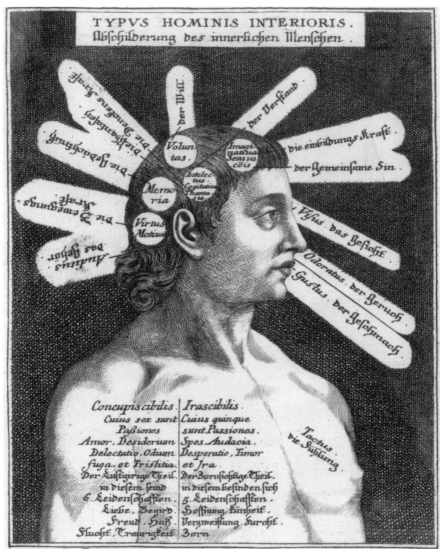

15

Anonymous

*Image of the Inner Man*

1779

from *Ichnographia Emblematica Triplicis*

16

Caspar David Friedrich

*Morning in the Riesengebirge*

1810–1811

Anton von Maron

*Johann Joachim Winckelmann*

1768

and estranged romantic works, inspired by Ludwig Gotthard Kosegarten's shore sermons,[44] are fully commensurate with a Winckelmannian neoclassicism grounded in a rhetoric of physical absence.

The German art historian's Pietist call for spiritual renewal urged the control of fleshy imagery through abstract literacy (fig. 17). Classicism and reading together were to vanquish the tyrannical "Popery" promulgated by Bernini's hyperbaroque art. It was not accidental that Winckelmann's "Dissenting" aesthetics touted the simplicity, sobriety, and measure of the fastidious ancients over the invidious and despotic mod-

erns.[45] Significantly, this connoisseur and critic felt most at ease when arousing corporeal properties were filtered through the cool medium of reproductive engraving.

The strange tension between frugality and sensuality exhibited in Winckelmann's writings about ancient art derived from the Puritanical temptation by, and rejection of, desire. Reading and writing facilitated the sin-free return to the same sources without the audience's being subjected to the perils of tactile color and whorish paint (fig. 18). Similarly, seductive archetypes of Greek sculpture, such as the *Antinoös,* could be transformed into chaste objects of religious meditation. Exquisitely isolated and safely distanced from the body of the perceiver through the mechanical intervention of the printmaker, Hellenistic statues had their erotic physicality tempered. Hallowed works of art thus resembled devotional texts. They, too, were available for intimate study and veneration in private retreats withdrawn from the public eye.

In sum, we should think of the Protestant culture of reading and writing as a fundamental indictment leveled against a supposedly Catholic fetishism. Defined by the French mythographer Jacques-Antoine Dulaure as more primitive than idolatry, fetishism was practiced prior to image worship. It was the earliest religious cult and thrived among the least developed nations. Coinciding with the barbarous infancy of society, it was the reification of "everything that makes a strong impression on the senses."[46] Thus natural and artificial phenomena were designated as cherished objects, desirable goods, to be worshiped because they housed occult powers. Rousseau's diatribe against the corruptions of civilization, I believe, can be understood as a jeremiad against the "Oriental" fetishization of the marvel-mongering arts and sciences. Their "despotic" development facilitated the birth of a modern and credulous consumer culture.

Rousseau's dictum that the literature and learning of the eighteenth century tended more to destruction than to edification[47] was surely based on his firsthand experience of the regnant Jesuit method of inculcation. In the *collège,* persuasion was achieved through a succession of aestheticized experiences. His shocking antiprogressivist stance was subjected to an immediate and vigorous attack. One notable critique mounted against the great *solitaire* was by Augustin Roux. The French physician and chemist excoriated the author of *Emile* (1762) for not having seen that ignorance, rather than knowledge, was the enemy. Like Rousseau, Roux operated outside the ambience of the *Encyclopédie,* even publishing a counter *Nouvelle encyclopédie* (1766). He became the apologist for the concrete, practical, and nondelusory sciences so conspicuously absent in earlier centuries marked by superstition and slavery. Instead of denouncing technology, Rousseau should have exposed those impostors who abused the credulousness of the masses and erected "ridiculous gods." Confronted by rational men, these "horrors" would have dissipated like nocturnal phantoms melting in the first rays of the sun.[48]

Rousseau's intrusive mode of training, vigilantly alert to the blandishments of sophisticated knowledge, was not exceptional.[49] George Turnbull, Professor of Moral Philosophy at Aberdeen and an important theorist of early learning, asserted the importance of teaching the young how to make ethical distinctions. Nothing was more valuable than the ability to separate useful inquiries from "those that ought only to have the place of Amusements, like a Game at Chess or Picquet."[50] Unlike Rousseau, however, and in typically Scottish Enlightenment fashion, he emphasized the empirical and inductive sciences. Nonetheless, Turnbull stressed that along with the "polite arts," the sciences should be supported only if they functioned as moral "Examples" or "Experiments in Philosophy." "Public Ornaments," not "private Furniture," the arts and sciences must promote virtue, not feed contradictory passions and appetites. Rousseau's detestation of insincerity was matched by Turnbull's hatred of feigning. The Scot singled out the educational system of ancient Athens. In that golden age "no Hoodwinking or blinding Arts" were used, although vice was represented in its true colors.[51] In the new moral climate of the mid-eighteenth century, the follies and corruptions of humanity had to be painted veristically and not presented as a captivating spectacle.

### Ingenious Pastimes

For the high enlighteners, mental release signified the absence of critical thought characteristic of hegemonic theocracies from antiquity to the present. Intellectual "vacation" was also synonymous with the apparent lack of work that went on in vision. Mere beholding was contrasted to the exertion of reading and writing. In addition, there was another connotation lying beyond these two and embracing them. This third concept played a major role in the intellectual and social history of the period. The phrase "the mind's release" also referred to that temporary breathing of the soul occurring during an interval of leisure. Cognitive drift left a void needing to be occupied. Rolicking pastimes (fig. 19), innocent and not-so-innocent sensual pleasures (fig. 20), outdoor (fig. 21) and indoor (fig. 22) spectacles, aristocratically feminine masquerades (fig. 23), and gentlemanly blood sports (fig. 24) vied with ingenious optical or geometrical "problems" (fig. 26) and "mathematical recreations" to while away the empty hours.

By definition, the soul's vacancy was inseparable from useless vanities (fig. 25). Soap-bubble blowing children routinely symbolized the fruitlessness

19

David Vinckboons

*Mountain Landscape with Merry Company*

17th century

20

Jean-Honoré Fragonard

*The Swing*

1768

21

Francesco Guardi

*Balloon Ascent*

1784

22

Philibert-Louis Debucourt (?)

*The Dressing Room of the Extras*

*of the Comédie Française*

c. 1785

23

Jean-François de Troy

*La Toilette pour le bal*

1735

24

George Stubbs

*Captain Samuel Sharpe Pocklington*

*with His Wife, Pleasance, and (?) His Sister Frances*

1769

of filling in the blank. Whether stationed beneath the highest creations of human reason, as in Sebastien LeClerc's geometrically derived lines, or dwarfed by the sumptuous products of the arts and sciences, as in Cornelis de Vos's allegory, no material object could stop up the spiritual gap.

The long twilight of the baroque emblem tradition stretched well into the eighteenth century. It carried with it that earlier era's dark intimation of mortality and existential fragility. Although transformed by the fashionable physicotheology of the Deists, the batlike fly in Johann Jakob Scheuchzer's encyclopedic *Physique sacrée* (1732–1737) still hieroglyphically signified transitoriness and nothingness. The scientific specimen remained pinned above the hollow man of dust and the evanescent toys of youth (fig. 27).

I want to examine certain high points in the intellectual trajectory arcing from innocent "filler" amusements to transformative mathematical recreations. Evolving from the witty and hermetic conceits originating within the Jesuit society of the high baroque, these technological games were altered by the eighteenth-century culture of curiosity. Aristocratic *jeux d'esprit* and occult *problèmes divertissans* had to accommodate a new public and private ornamental science mania. Characteristic of the Enlightenment, such popular demonstrations were founded on the principle of educating while entertaining. Part of the mass literacy movement initiated during the *ancien régime,* didactic visual performances increasingly set themselves in opposition to the abstruse, not-for-profane-eyes rhetoric of late sixteenth- and seventeenth-century treatises.

The psychological anthropologist Roger Caillois divided the world's games into four groups: *agon,* or those in which competition is the main feature; *alea,* or all games of chance; *ilinx,* or those creating vertigo by scrambling ordinary perceptions; and mimicry, or activities in which alternative realities are created.[52] According to this classification, baroque science games essentially represent the merger of *ilinx* with mimicry. Esoteric optical entertainment, transmitted in an eccentric illusionistic style, did not really teach but decorated the surface of privileged leisure.[53] In the eyes of the seventeenth and eighteenth centuries, the mannerist *jongleur* Giovanni Battista della Porta was the undisputed originator of *scherzare.* Thus Henri-Gabriel Duchesne, an important advocate for French industry, stressed the distinctively modern alliance between purposeful technology and playful intellectual constructions in the Neapolitan ma-

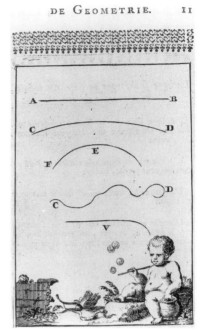

25

Sebastien LeClerc

*Definition of a Line*

1691

from *Pratique de la géométrie*

26

Cornelis de Vos

*Allegory of Transitoriness*

17th century

TAB. DXLIX.

PSAL. XXXIX. v. 12.
Tinea, pulvis et umbra.

Psal. XXXIX. v. 12.
Der Mensch eine Schale und Eitelkeit.

I. A. Fridrich sculps.

27

Johann Jakob Scheuchzer

*Vanitas*

1732–1737

from *Physique sacrée*

28

Giovanni Battista della Porta

*Burning Glasses*

1658

from *Natural Magick*

gus's creation of learned wonders. Duchesne, in his *Notice historique sur della Porta* (1801), claimed that the Italian polymath was unfairly known in his day only as the inventor of the telescope and the camera obscura rather than as an experimentalist who developed over 1,350 techniques.

Duchesne's intellectual biography turned della Porta into a precursor of the Enlightenment. Thus the reader was informed how the inventor struggled in an era when the printing press was just beginning to disseminate information and the sciences "were still covered with the rust of ages."[54] *Natural Magick,* one of the great uncanonical texts of the scientific revolution, was first published in 1558 and went through many editions in the following two centuries. This small manual contained veiled descriptions, having deliberately been "written obscurely."[55] The text, interspersed with woodcuts, spread his fame throughout Europe as the inspired creator of phantasmagoria. Singular optical instruments, such as "Burning Glasses, and the wonderful Sights to be seen by them" (fig. 28), were given pride of place. These clever devices situated him within that great underground current of the technological marvelous flowing from the eccentricities of mannerism, to the anamorphoses of the baroque, to the automatism of romanticism.

Della Porta's slim volume occupies a large niche in the history of mathematical recreations. Clear and succinct illustrations conveyed information at odds with the mystifying riddles he felt obliged to employ so as not to get into trouble with the Inquisition. His preface made plain that the manual was intended for the "ingenious Reader." Studious types possessing "wit," who thoroughly researched the topics, might aspire to unravel their meaning. Beyond fear of religious persecution lurked an obvious aristocratic elitism. "My catalogue of Rarities," he proudly asserted, was not destined for "the publick View of all Men." There were many excellent things fit only for "the worthiest Nobles which should ignorant Men that were never bred up in the sacred Principles of Philosophy come to know, they would grow contemptible, and be undervalued; as Plato saith to Dionysius, they seem to make Philosophy ridiculous, who endeavour to prostitute her Excellence to prophane and illiterate Men."[56] In spite of such arcane exclusivity, della Porta was an important precursor for a democratizing experimental science. Verbally, at least, he separated "wicked Curiosity" and "black Magic," or the unnatural artifices of sorcery, from the proper "Survey of the whole course of Nature." He demonstrated for subsequent generations that the practical part of natural philosophy consisted in manufacturing licit effects.

Lawfulness and ingenuity were the constantly touted properties of those cognitive exercises meant to restore the depleted mind. "Recreation," according to William Leybourne's *Pleasure with Profit* (1694), was a second creation. "When weariness hath almost annihilated our *Spirits:* It is the breathing of the Soul, which would otherwise be stifled with continual Business." This "Interval" of "Rest or Recreation" allowed the body "to recollect its pores, gather its breath to fall fresh again into its first career."[57]

Leisure, as nonessential activity, temporarily plugged those gaping holes opening during the long day's journey into night. Spots of time not given over to gainful employment could profitably be filled with cultural pastimes. Della Porta's model of a hermetic private academy in the home quickly evolved into the no longer exclusively high-class consumption of useful novelties.[58] Paris and London, the geographical and social centers of fashion in the early seventeenth century, witnessed an explosion in the manufacturing and buying of toys, mirrors, cards, puppets, but above all of illustrated and utility-justified "mathematical recreations" (fig. 29). These encyclopedic grab bags of conjuring tricks, amazing feats with coins, dice, and cards, "experiments" with chemical and optical apparatus, mathematical puzzles, and "ingenious" problems in hydrostatics and mechanics were obvious compilations rapidly spreading know-how from capital cities into the provinces. The piecemeal method of gathering unusual articles or extraordinary recipes into a sort of two-dimensional cabinet of rarities was visually captured in the inclusion of a mosaic composed of postage stamp diagrams. Such compartmentalized miniatures ornamented one of the earliest and most frequently cited examples of "useful and recreative" works.[59]

The baffling Henry van Etten's prototypical *Mathematical Recreations* proved, however, to be a palimpsest. Trevor Hall, who sorted out the knotty problem of priority, discovered that the 1633 London edition was itself merely the translation of an earlier, brief book published in Lorraine in 1624. The Rouen edition of 1628 crudely simplified the original and graceful engravings (tipped into the text) by the Lorraine artist-printer Jean-Appier Hanzelet. The subsequent schematic representations by Henrion and Mydorge found their way into later editions, such as the one by van Etten. The Rouen publication also separated the treatise on *feux de joie,* or artificial fireworks used for amusement, from the patchwork of 91 problems (fig. 30).[60]

29

Henry van Etten

title page to *Mathematicall Recreations*

1677

30

Henry van Etten

*Artificial Fireworks*

1677

from *Mathematicall Recreations*

This concluding essay on pyrotechnics, entitled "Of Recreative Fires," was typical of the ungainly mixture characterizing all the entries. It described how the would-be *artificier* might design, construct, and execute "deceitful Candles." A fiery dragon was the most charming of these. Guaranteed to make the loudest noise and elicit the greatest astonishment, the mechanism was remarkable for the addition "of much grace to the action." This witty beast—composed of gunpowder and tallow and mounted on rockets—ran up and down a rope chasing combatants similarly spewing sparks from the packed cavities of their bodies.[61] Van Etten's impractical and unreproducible toy was the comical antithesis of the French engineer Bernard Forest de Bélidor's elegant *bombes d'artifice* (fig. 31). Flaming blazons, portraits, ciphers, and symbols of the Sun King were an integral part of the festival machinery destined to amuse and impress the court. In addition to providing illustrations of entertaining emblems and political messages, Bélidor's practical treatise was usefully accompanied by meticulous instructions visually demonstrating the actual manufacture of fireworks (fig. 32).[62] In this aleatory culture, it was apt that playing cards formed the cylinders used to encase explosives before their insertion into molds.

31

Bernard Forest de Bélidor

*How to Make Suns and Letters Appear in Flames*

1734

from *Le Bombardier françois*

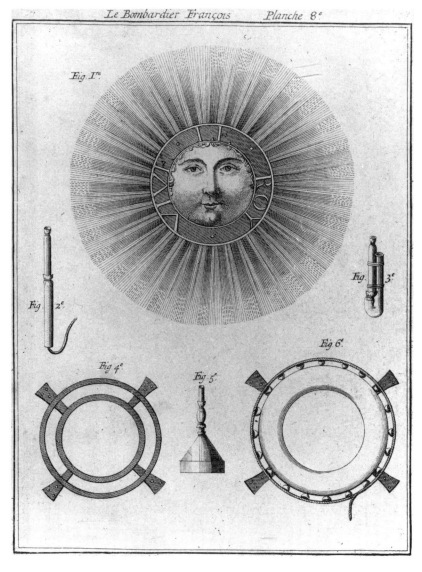

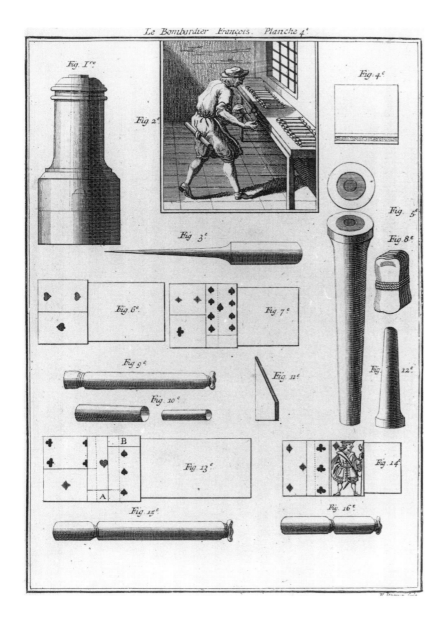

## 32

Bernard Forest de Bélidor

*Demonstration of How to Make Rockets*

1734

from *Le Bombardier françois*

But we must return to the mysterious van Etten. Significantly, he was a pupil of the Jesuit Jean Leurechon, Professor of Mathematics, Philosophy, and Metaphysics at the University of Pont-à-Mousson.[63] The Jesuit *collège*, with its emphasis on sensory spectacle concealing functional rope and pulley, was unmistakable in the Dutch author's preface. Van Etten explained how he was drawn into "the search for demonstrations more difficult and serious" to "delight" the nobility and gentry. "Rare and curious propositions," he emphasized, were not pursued "in the hope of gaine to fill their *Purses*."[64] "Admirable experiments," then, were to provide purely psychological profit. Like diverting firecrackers, they offered decorative visions untainted by tradesmanly drudgery for money.

This financial distinction is critical. In the course of the following two chapters, we shall see how the shifting boundaries between amateur and specialist, unpaid and paid demonstrator, were linked to battles over proper or improper social status. For the moment, it is important to register the proliferation of such virtuous sports conducted "upon severall things of small consequence, as upon the foote of a fly, upon a straw, upon a point, nay upon nothing; striving as it were to shew the greatnesse of their glory in the smallnesse of the subject . . . to solace the mind and recreate the spirits."[65]

"Heaping together" diverting experiments in a *Wunderkammer* manner was consonant with the fashionable demand for information presented in sugar-coated disguise. Thus Claude-Gaspar Bachet de Méziriac's *Problèmes plaisans* (1611) was a small treatise of amassed games. Although purportedly stocked with "bagatelles, & choses du tout inutiles," he maintained that these amusing arithmetical and geometrical problems were visual exercises in simulation. Just as a brave captain does not put his army at risk without first training soldiers through imaginary drills, so too the speculative sciences aid merchants, architects, and engineers in the art of hypothesis or *Gedanken* experiments.[66]

A hallmark of the baroque "art must conceal art" aesthetics was the premium placed on dexterity. A nimble *façon de faire* ravished the beholder's eye without revealing the mechanism behind the artifice. Thus Thomas Johnson's *New Booke of New Conceits* (1630) admonished its readers to practice useful "tricks" subtly.[67] Further, Nicolas Hunt's rare and uncharacteristically substantial *Newe Recreations* (1633) suggested a compelling analogy of games to life. Their practice during interludes or free time not only prevented the mind's dangerous lapse into lethargy or chaos, but provided a harmless model for how to increase productivity once one returned to the commerce of living. Yet beneath the rhetoric of sensory delight in skilled performance lurked the real fear that one had to develop attention to avoid anomie: "Whet the wits of indisposed Sluggs and stirre up dull inventions, and melancholy dispositions to a laudable exercise of their vacancie, and sequestration from serious imployments."[68]

No one did more to establish an entertaining and instructive industry of light and magic than Athanasius Kircher.[69] This Jesuit from Fulda, and his colleagues and successors at the Collegio Romano, ushered in a new era of folio and even multivolume encyclopedias. A specialist in technological *concetti,* Kircher created metaphor machines for the display of fantastic images in a museum of paradoxes.[70] Beguiling mixed-media experiments in optics, hydraulics, automation, and acoustics were disseminated throughout Europe in publications by himself and numerous colleagues and disciples. These marvelous toys were designed to provoke *stupore* by improving upon the arcane inventions of della Porta. Rotating cylinders, wired levers, and driving pumps were similarly concealed from the startled and delighted beholder (fig. 33). Automata such as the nymph Echo, the piping Pan, and a crowing and wing-flapping rooster were remotely set in motion by the ingenious *phonotacticum* revolving in a subterranean grotto.[71]

Kaspar Schott's *Mechanica Hydraulica-Pneumamatica* (1656) and the wittily entitled compendium of scientific games, the *Joco-Seriorum Naturae et Artis* (1664), were composed after his return to Germany. In Rome, he had visited and studied with the celebrated Kircher. The seventeenth century was remarkable for the creation of magnificent frontispieces, hence the architectural entry to Schott's compendium of natural magic matched the prodigies it housed (fig. 34). Hercules and Hermes, in their paired classical niches, emblematically incarnated the intellectual strength and mental agility acquired by the *virtuosi* who consulted this thesaurus. Having lived through the Thirty Years' War, Schott crowded the broken trophies of battle into the attic story. Instruction and peaceful play were thus shown to be inseparable. Moreover, the German Jesuit sounded a theme that would be repeated with many variations during the Enlightenment. Such "Physica ac Mathematica Ludicra ac Recreationis" were not intended merely to delight but to help spectators guard against fraud.[72] As he had painfully experienced, machines could become machinations in the wrong hands.

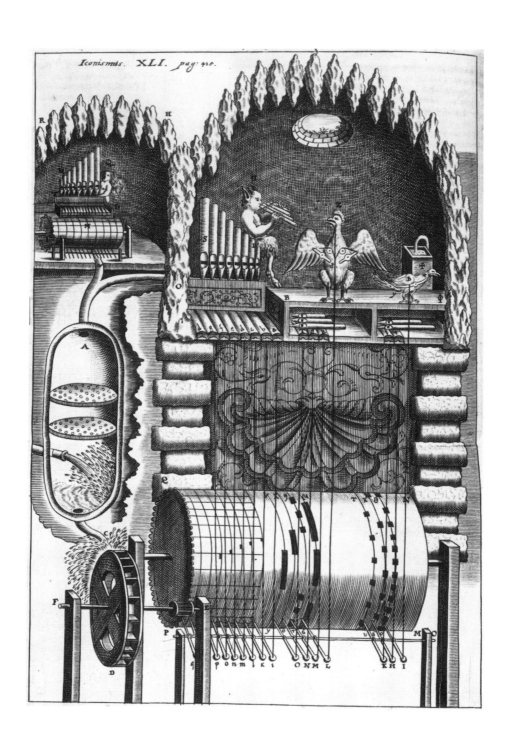

33

Kaspar Schott

*Water Organ, Piping Pan, and Rooster Automata in a Grotto*

1656

from *Mechanica Hydraulica-Pneumatica*

34

Kaspar Schott

*Hercules and Hermes Flanking the Portal to Knowledge*

1664

from *Joco-Seriorum Naturae et Artis*

Schott's desire to provide a literate audience with a gamut of pleasurable employments, stretching harmoniously from the serious to the jocose, had an important precursor in Daniel Schwenter's *Deliciae Physico-Mathematicae*. This handsome volume, published posthumously in 1636, was addressed to all ages but especially to the young (fig. 35). Immensely popular, the book was enlarged by two supplementary editions appearing in 1651 and 1653. Professor of Hebrew, Oriental Languages, and Mathematics, this Jesuit compiled a comprehensive collection of recreational experiments based, we now know, on van Etten's work, but augmented by additional problems.[73] Importantly, van Etten's crude, juxtapositional mosaic was metamorphosed on the bilingual title page into a finely executed and flowing garland of select *Erquickstunden,* that is, recreations. In this philosophical *margarita,* or daisy chain of knowledge, schematic geometric figures logically evolved into illusionistic miniatures. These emblems instantiated architecture, navigation, astronomy, and even swimming and tightrope walking. Significantly, Schwenter complained of the (to him still unknown) author of the *Mathematical Recreations'* reliance on cards and dice to teach scientific principles.

This scrupulousness and casuistic nicety, unknown to his forebear, was probably a response to a more stringent Protestant morality. Schwenter deliberately eschewed the worldly tokens of gambling when he devised children's games. Indeed, for him, Bonanni, and a host of later Jesuit educators such as the Abbé La Pluche, natural history with its "innocent pleasures" became the virtuous discipline par excellence. According to these devisers of an aestheticized ethics, even blowing bubbles (see figs. 25, 26, 27) was not a trivial activity when the child pondered why the tiny globes were round and not some other shape.[74]

Georg Philipp Harsdörffer (1607–1658), the Nuremberg poet and literary theorist, was among the most distinguished baroque creators of serious amusements. Famous for his *Klangmalerei* or tonal painting, he is less known as the author of a three-volume *Delitiae Mathematicae* (1651) (fig. 36). The monumental broken-pedimented portico of the title page shows regal mathematics personified as wisdom. The winged head betokens elevated thought. The scepter she holds and the diagrammatic globe inscribed on the immutable cube upon which she stands symbolize her dominion over the arts and sciences. Naked putti, swarming over the building, handle instruments and otherwise engage in scientific sport. These are the children and the young at heart who learn through recreation.

35

Daniel Schwenter

title page to *Deliciae Physico-Mathematicae*

1651

36

Georg Philipp Harsdörffer

title page to *Delitiae Mathematicae et Physicae*

1651

Significantly, Harsdörffer visually juxtaposed *Kinderspiele* (mirror, whirligig, and inflated bladder) with *Künstlerspiele* (abacus, armillary sphere, sextant, calipers, and other writing and measuring implements). From the attentive and observant expressions registered on the infants' faces, it is apparent that both toys and apparatus foster understanding. In fact, Harsdörffer hoped schoolmasters would use his book to instruct beginning learners pleasantly.[75]

Like many of his baroque sources, Harsdörffer's compilation was marked by the period's fascination with the art of visual illusion. Citing the examples of Mersenne, Descartes, and of course Kircher, he posed the puzzling problem common to all books on natural magic. How could a knowledge of optics make objects appear and disappear? The camera obscura was the chief instrument allowing the early moderns to catch a fleeting glimpse into the normally hidden operations of perception (fig. 37). Harsdörffer's psychologically disorienting prank (*ilinx*) stimulated viewer pleasure through bafflement. The beholder was perplexed when confronting a set of nesting optical boxes. Within the confines of a camera obscura, ABCD, lodged another, smaller and translucent box, EF, constructed of oiled paper. A mirror placed at the aperture, F, introduced a train of edifying and scary specters, such as the skeleton holding a scythe. These apparitions were reflected on the bottom of the portable trick cabinet but might also be projected on a wall,[76] as in the Premonstratensian monk Johann Zahn's magic slide show (fig. 38).[77]

Anamorphoses, ambiguous figures, and depictions of impossible objects intrigued Jacques Ozanam (1640–1717), whose *Récréations mathématiques* (1692) were to become the most influential French example of this genre for the Age of Reason. This prolific and serious mathematician was a member of the Académie des Sciences and thus departs from the profile we have been sketching. A Jew, he converted early to Christianity. Repelled at the thought of being destined for the priesthood, he became a private tutor in mathematics instead. Working at Lyons and then Paris, Ozanam gained wide acclaim for his scholarly treatises. He enters our story because of the frequently reprinted and emended editions of the *Récréations*.

Ozanam freely acknowledged his debt to the long tradition of formulating *problèmes diverstissans* stretching back, in the case of the moderns, at least to Bachet de Méziriac and Kircher. He claimed that his *trompe-l'oeil* never

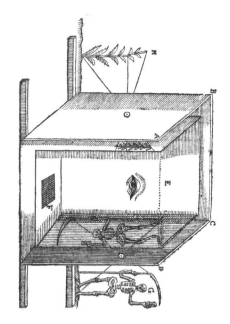

37

Georg Phillipp Harsdörffer

*Camera Obscura with Projection*

*of a Skeleton Holding a Scythe*

1651

from *Delitiae Mathematicae et Physicae*

38

Johann Zahn

*Magic Lantern Projections*

1686

misled, even momentarily, by allowing the seemingly solid world to shift for an instant (fig. 39). Consequently, this form of mental relaxation was genuinely new. It represented a specifically enlightening form of entertainment, worthy not only of children but of statesmen. The preface to the first edition traced the lineage of *jeux d'esprit* back to the Chaldean and Egyptian astronomers. These ancient wise men, wishing to attract the young to difficult studies, did so by arousing the "pleasure of curiosity."[78] The necessarily gymnastic and eccentric witness to Jesuit visions—seizing hermetic or religious scenes obliquely—was transformed by Ozanam into a rational observer sagely confronting the world head-on. His curiously centric anamorphic eye, unlike Harsdörffer's marginalized skeleton, no longer required the beholder to approach the object sideways.[79] The unambiguous revelation of the method whereby the image was thrown off center and distorted endeared his work to a future generation of *philosophes.* They, too, hunted after the calculable point of intellectual resolution in the midst of perceptual confusion.[80]

### Rational Recreations

It is difficult to pinpoint precisely the moment when gallant games and ingenious mathematical recreations became "philosophical and rational." Children's literature derived from the push toward alphabetization, which in turn created a literate public.[81] Reading aloud in the salon or home (fig. 18), with its conversational and theatrical style, was not supplanted by the private, silent, and solitary ingestion of abstracted information until late in the eighteenth century (see figs. 14, 17).[82] Such imagistic performance made learning, like food, palatable. Participatory enactment, I suggest, was central to the aim of rational recreation. It made abstractions concrete by picturing the practices of science. Material was internalized interactively. From the stern perspective of the Enlightenment, knowing speech managed to avoid the twin dangers of passive spectacle, fostered by the Jesuits (fig. 8), and eroticized epistolary exchanges, incited by indolent women (fig. 40).

Recall that the Jesuit system, widely operative in France until their expulsion in the 1760s, was predicated on an aestheticized *physique des enfants.* As the Abbé La Pluche declared in his epoch-making *Spectacle de la nature* (1732–1750), taking the amusing or gentler path to learning was useful. Even at the most tender age, the child should be encouraged to look "solely upon the exterior [of nature] or at that which strikes the

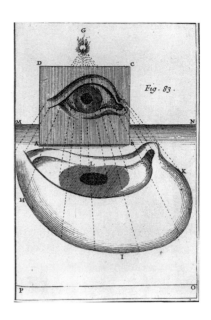

39

Jacques Ozanam

*How to Describe on a Horizontal Plane*

*a Deformed Figure Which Appears*

*Correct on a Vertical Plane*

1694

from *Récréations mathématiques*

40

François Boucher

*The Discreet Messenger*

18th century

senses." Penetration of the universe's secrets was a "delving task" better left to a superior intellect.[83] Rousseau's *Emile* (1762), infused by Jesuit pedagogical ideals, similarly asserted that the preceptor needed to make the pupil's ideas correspond to his sensations. The masterpiece of education certainly was not the reasonable man, as Locke had stipulated. On the contrary, it was the morally sincere being who was conditioned through sight to imitate virtuous acts.[84] Even the Jansenist Charles Rollin's Augustinian-tinged pedagogics touted the winning graces of vivid examples, always more efficacious than dry precepts. Indeed, because of original sin, children came into the world "surrounded by a cloud of ignorance." Hence they must be enticed from their inherent propensity toward indolence, idle play, and debauchery.[85] The seductive charms of rhetoric thus prepared the heart, even during vacant hours, allowing instruction to enter easily.

While there was a real continuity in the aims and types of baroque mathematical and Enlightenment rational recreations, there was also a significant difference. This nuanced change, I think, can be addressed by contrasting undemanding spectacle, or a type of watching in which the intellectual labor was concealed, with overtly hard-working visual persuasion. John Locke's widely translated and frequently republished essay *Some Thoughts Concerning Education* (1693) provided the impulse for the transformation I am describing. From his firsthand experience as a private tutor, he knew that children hated to remain idle. Moreover, effective education—and in this he resembled the Jesuits—implied that the preceptor "must make what you would have them [the young] do a recreation to them and not a business."[86] Paradoxically, Locke even urged parents and educators to use insincerity, a stratagem that was to enrage Rousseau. Thus if teachers wanted to dissuade their pupils from wasting time at top and scourge rather than reading and studying globes, then they must command them to play until they sickened of these toys. The example of *supposedly* forbidden adult recreation was enjoined to arouse the child's desire to learn through emulation.[87]

Locke's illusionistic tactics of visual persuasion, founded on the imitation of adult behavior that only *appeared* to be play, may be seen in a crop of scientific demonstrations that were really child-rearing exercises. From the Abbé Nollet's literally electrifying lectures, performed in his technologically advanced *cabinet de physique* (fig. 41), to Abraham Trembley's exhibitions of the astounding recuperative properties of the fresh water polyp

41

Jean-Antoine Nollet

*Nollet Demonstrating in*

*His Cabinet de Physique*

1743

from *Leçons de physique expérimentale*

42

Abraham Trembley

*Abraham Trembley with Children*

*in His Laboratory*

1744

from *Mémoires pour servir à l'histoire*

*d'un genre de polypes*

(fig. 42), girls and boys were both the producers and consumers of a sensationalized knowledge. The frontispiece to Nollet's *Leçons de physique* (1743) embodied his pedagogical credo. Sometimes it was necessary to abandon the difficult language of physics in order that this science "might be put in the hands of all the world." Employment as the private tutor of the Dauphin taught him that abstract principles, which could not be learned without laborious application, slipped easily into the mind when interspersed "with interesting experiments." The young, in particular, were repelled by hard words. Yet the current *goût de la physique,* he argued, proved that its abstruse operations could be made concretely appealing in a visual medium.[88]

Standing apart from the merely ostentatious possession of expensive singularities (fig. 43), the routinized drudgery (fig. 44) or sullen apprenticeship (fig. 45) of child labor, and adult-dictated play-acting (fig. 46), rational recreations were a sort of joyful diligence. Instructive scientific games were existential rehearsals. They incarnated the engrossing and unselfconscious art of experimentation fundamental to the laboratory and in ordinary life (fig. 47). Both as instrumentalized performance and as illustrated guide to serious amusements, the genre phenomenalized instruction. It erased the dualism between mind and body, art and craft, science and technology. Image and speech, thought and activity were put in direct touch in the process of making a simultaneously intellectual and physical good.

43

Jean-Baptiste Perronneau

*Portrait of Bonaventure Journu*

1767

44

Johann Wilhelm Meil

*The Book Printer*

1765

from *Spectakulum Naturae & Artium*

Separator
Ein Klaube Jung

45

Christophe Weigel

*Separator*

1721

from *Abbildung und Beischreibung*

*deren sämtlichen Berg-Wercks*

46

Jean-Honoré Fragonard

*Boy as Pierrot*

c. 1785

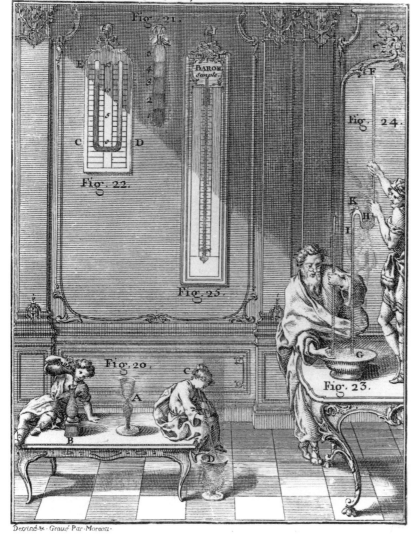

## 47

Jean-Antoine Nollet

*Demonstration of Different Weights*

*of Air and Water*

1743

from *Leçons de physique expérimentale*

The pleasures of science were taught by a growing number of private tutors and freelance lecturers. Like Nollet or Sigaud de La Fond (to be discussed in chapter 3), they catered to a now vanishing breed of wealthy amateurs. These dilettantes took childlike delight in the cultivation of hobbies that imparted order to their free time. The lure of recreation was consonant with Locke's and Rousseau's sensationalist view that the sense organs had to be continuously exercised.[89] This highly contemporary notion that play is fundamentally about activities with self-contained rewards is evident in the spate of publications claiming to be "philosophical amusements" or "easy and instructive recreations for young people."

These included Henri Decremps's collection of "curious experiments" engaging the attention by making information "familiar and entertaining."[90] The French geographer and physician Edmé Guyot's (1706–1786) eight-volume *Nouvelles récréations* (1772) offered a tantalizing array of apparatuses unfolding a succession of diverting appearances. Director of the French postal service, Guyot was passionately interested in improving the courier system to facilitate communication between merchants, industrialists, and manufacturers in far-flung localities. Believing that the expanding domain of knowledge also required organization, Guyot insisted it was the nature of the human mind to be active in its interaction with the external world. Since perception was not passive, reason might be deceived. Anamorphoses, concave mirrors, and ghost-projecting machines demonstrated that, of all the senses, sight was most prone to illusions (fig. 48).[91]

The fashionably updated and vastly expanded 1778 and 1790 editions of Jacques Ozanam's *Récréations mathématiques* amounted to a new work arguing for the marriage of the useful and the agreeable, the manual and the mental. In pointedly revisionist editorial remarks, Jean-Etienne Montucla (1725–1799) emphasized that mathematics was no longer a hermetic activity but integral to public education. Part of the circle of enlighteners including d'Alembert, Diderot, Cochin, Blondel, and Le Blon that gathered at the *libraire* Jombert, he proselytized on behalf of demystification through the concrete learning of optics, mechanics, astronomy, geometry, and physics. In his guise both as a journalist associated with the *Gazette de France* and as a historian of mathematics, Montucla criticized Ozanam's volumes because they were addressed to amateurs yet were filled with ingenious and recondite queries appropriate only to professional mathematicians. Such enigmas, he declared, were wholly inappropriate to the general public of the later eighteenth century. Van Etten's *Récréations mathématiques* came in for even harsher judgment as a "pitiful rhapsody," a "trashbin [*fatras*] of problems."[92]

Significantly, Montucla employed a style dramatically different from the nettlesome abstractions thronging Ozanam's 1692 edition. Young beginners, at the inception of their studies, needed to be piqued by Greuze-like visual narratives. A perceptive psychologist, Montucla devised poignant problems in practical mathematics. He gave the example of the father of a family who, on his deathbed, bequeathed a plot of land to his three sons. How was the inheritance to be divided? As an advocate for progress,

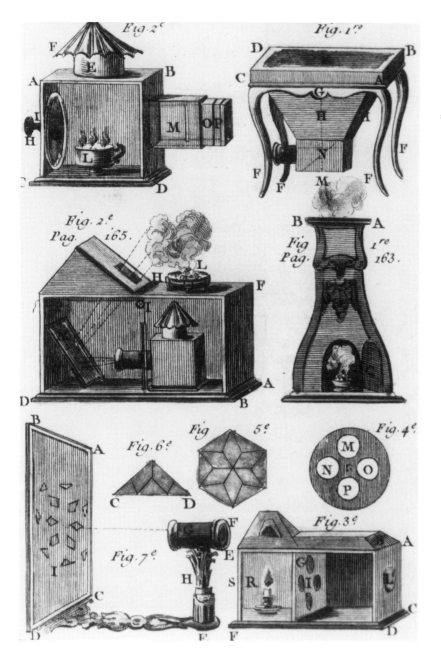

## 48

Edmé Guyot

*"Painting on Smoke"*

*with Magic Lantern and Prismatic Projections*

1772

from *Nouvelles récréations*

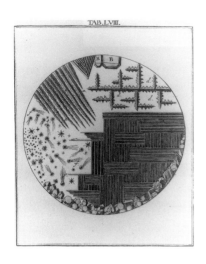

TAB.LVIII.

49

Martin Frobenius Ledermüller

*Alum Crystals*

1776

from *Amusemens microscopiques*

Montucla also deemed Ozanam's publication to be woefully out of date. Jacques Lacombe's *Dictionnaire encyclopédique des amusemens* (1792) similarly commented on the recent proliferation of inventions and the betterment of older technologies. Magnetic, electrical, and chemical discoveries abounded, automata were being improved, and microscopy continued to be perfected (fig. 49).[93]

Martin Frobenius Ledermüller's collection of microscopical amusements was published serially in German, Dutch, and French editions between 1760 and 1776. The success of this venture—long on illustrations and short on letterpress—attested that even specialized recreations could be profitable. Butterfly wings, silk cloth (directed particularly at women readers), flint sparks, sea sand, and crystals of all sorts were magnified to reveal patterns of shapes and colors captured in delicate hand-colored engravings.[94]

William Hooper's four-volume *Rational Recreations* borrowed many experiments and illustrations from Guyot's work. This venture, like that of his French precursor, was indicative of the later eighteenth-century tendency to publish in multiples. It was also a typically late Enlightenment production in its emphasis on optical illusion as delusion. Hooper's claim to rationality for his *Recreations* was based, characteristically, on their anti-"despotic" goal. The aim of these pleasing problems was "to show the fallacy of what men regard as most certain: the evidence of the senses." "Instructive entertainment," thus, had become inseparable from the difficult task of rendering useful knowledge engaging and nondeceitful. It was "the rising generation" as well as the jaded adult for whom this improbable data factory was created. Filled with machines for representing nebulous phantoms, magician's mirrors, and cinematic boxes, illustrated popular science books released a flow of modish two-dimensional images to be constructed or erected in the teaching laboratory of the home (fig. 50). Learning became the seductive coercion of perception. Part of a burgeoning leisure industry, information was "not dull, tedious, and disgustful, not rugged and perplexing, not austere and imperious, but facile, bland, delightful, alluring, captivating."[95]

### *"The Boy's [and Girl's] Philosophy"*

Playing with ideas was an adult method for disciplining wayward childhood activities. Yet rational recreations occasionally tried to avoid the

50

Charles-Amédée-Philippe van Loo

*The Magic Lantern*

1764

semblance of parental control by having children teach children (fig. 51). Emulating the much-quoted and translated Locke, educators writing in the 1740s urged tutors to be childlike, to "come down to [the student's] capacity as much as possible." But it was preferable for teachers actually to resemble Chardin's perfectly groomed adolescent. Patiently guiding her small charge's halting progress through the alphabet, she turned learning into performance by the expert direction of a knitting needle. Rousseau later claimed that a good governor should be young, as young as a wise man can be or, better, a child himself.[96]

No one understood more astutely the fine art of "coming down" than John Newbery. Capitalizing on the market for illustrated books with wide margins, large type, concrete language, and compact enough to fit into still growing fingers, he wrote and published works refracted through children's eyes. "Tom Telescope's" tiny *Newtonian System of Philosophy* (1761) was probably composed by Newbery himself (although Oliver Goldsmith has also been proposed as the author).[97] Going through ten editions by the end of the eighteenth century, this philosophical conversation was by no means a unique book. The action took place at holiday time when young ladies and gentlemen gathered at the house of the Countess of Twilight. Debates ensued over how the "little gentry" were to divert themselves. One by one, hot cockles, shuttlecock, threading-the-needle, and blindman's buff were rejected. Master Telescope, a "young gentleman of distinction," sat silently until one member proposed playing cards for money. Passionately calling for innocent amusements, he condemned Bath. There, "a young *urchin* just breeched" or a little "doddle-my-lady in hanging sleeves" was led to a gaming table to "play and bet for schillings, crowns, and perhaps guineas among a circle of sharpers."

The high-minded and decidedly middle-class Master Tom opposed not just idle gambling but all amusements that did not "improve the understanding." He seemed to be indicting those undemanding pastimes engaged in by indolent aristocratic children idling away the hours in the fancy portraits of Reynolds and Gainsborough (fig. 52).[98] Conversely "our little philosopher," after having acquired the necessary instruments from the Marquis of Setstar, offered to teach the useful secrets of natural philosophy. Courses delivered "in the entertaining manner" were to stimulate a Lockean "train of pleasing ideas" in the audience. Repairing to a small parlor with tots and elders in tow, Tom whipped a top to demonstrate Newton's laws governing matter in motion. On the contrary, when

51

Jean-Baptiste-Siméon Chardin

*The Young Governess*

c. 1739

52

Thomas Gainsborough

*Marsham Children*

1787

Emile whipped his top Rousseau claimed that eye and arm were strengthened by the physical exercise, but that his protégé learned nothing.[99]

Pleasantly wedged between dining on "tarts, sweetmeats, syllabubs, and such other dainties as his Lordship thought were most proper to youth," Tom's experiments advanced in difficulty. Hoops, marbles, cricket bats, and other objects from ordinary life were drawn upon to make concrete points about the workings of gravity and friction. By the time the lecture-performances reached the topic of the senses, "the *Boy's* Philosophy" was addressed more to the adults in the company than to the young ladies and gentlemen. A lad, discovering himself in the glass (fig. 53), personified this picture book method of familiarizing science by turning it into a domestic adventure. The Lockean vision of concept formation was acted out by a child who, thinking he saw a new playmate, cried out "little boy, little boy." Eventually, and by the reaction of others to this spectacle, the idea dawned that something new had been acquired through sight. This empiricist epiphany came about when parents and nurse arrived on the scene to share in the diversion. Learning, as in the epistemology of Locke and Berkeley, began only when the infant felt his father's hand on his head and beheld the reflection in the mirror, which, in turn, was recorded in the storehouse of memory.[100] In sum, Tom Telescope's demonstrations of rational philosophy exemplified one way in which the picture book format slowed down, simplified, and socialized abstractions by representing them in more comprehensible visual terms.[101]

The gendered character of eighteenth-century natural history instruction is evident in the differing depictions of boys and girls found in these attractive primers. One of the advantages of the postmodernist critique is its promise of drawing young women into the sciences, mathematics, and engineering.[102] Thus while Tom exhibited heroic mastery over his material possessions and dominion over his female and male playmates, Benjamin Martin expressed a more egalitarian "standpoint" theory. Women's experience was also drawn upon to shape research practice (fig. 54).[103] As Nicolas Hunt's *Newe Recreations* earlier indicated, young "lords and ladies" were to be the principal consumers of gadgetized "sports" in this economy of learning that mixed sensory pleasure with intellectual profit.[104] The question remained, however, what were their respective roles to be? Recall that Emile was described as a victim of women whose passions and caprices buffeted the first seven years of his life. Rousseau identified feminine instruction in the home either with the burdening of memory

## 53

Tom Telescope

*A Boy Finds Himself in the Glass*

1761

from *The Newtonian System of Philosophy*

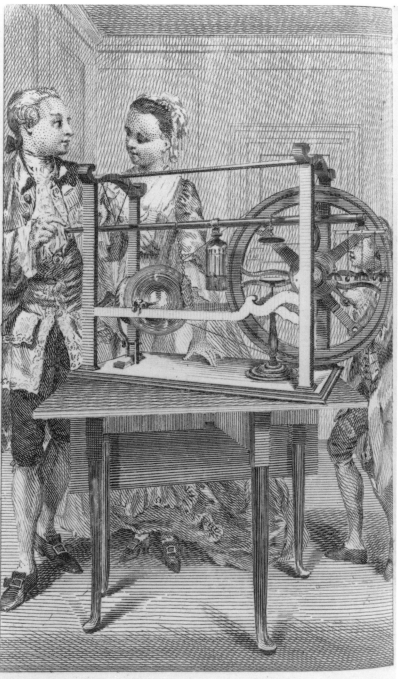

A New ELECTRICAL MACHINE *for the* TABLE

54

Benjamin Martin

*A New Electrical Machine for the Table*

1755

from *The General Magazine*

through the useless gathering and repetition of rote information or with frivolous games.[105] On the contrary, the successful instrument maker, itinerant popular science lecturer, and publisher Benjamin Martin relied upon the powerful casuistic strategy of individual example. Both sexes were to be attracted into the sciences by encouraging imitation.[106]

Martin's *General Magazine* was a complete library of the arts and sciences published over a ten-year span and intended to be "of easy access." Astronomy, pneumatics, optics, hydrostatics, hydraulics, and electricity were communicated in a direct and plain style to the children of the gentry. His engaging protagonists were a young undergraduate on vacation and his sister. Significantly, this educational entrepreneur avoided the standard format of a miscellaneous compendium that collected things "piecemeal, in bits and scraps, disjointed and mangled, without order or connection and therefore useless." Monthly installments, costing twelve shillings an issue, were to retail subjects in digestible and manageable portions.

There were *two* stars in this Fontenelle-inspired dialogue. A brother and sister, Cleonicus and Euphrosyne, embodied the fact that "it is now growing into a Fashion for the Ladies [also] to study Philosophy." Euphrosyne, unlike Fontenelle's marquise in the *Entretiens sur la pluralité des mondes* (1686), tended toward introspection. Image-conscious, she wished that it did not "look quite so masculine for a woman to talk of Philosophy in Company." Anxiety at potential peer disapproval led her often to sit silently "for fear of being thought assuming or impertinent."[107] It is important, then, that although Cleonicus commandingly puts his finger near the electrical fluid emitted by "a large and elegant Machine," Euphrosyne has risen to join him in the conduct of "a few entertaining Experiments." On one hand, while Martin generally limited Euphrosyne to asking questions of her expert brother, on the other hand, she was described Prometheus-like as occasionally holding the phial. Whatever the motive, and even if it were only to encourage the sale of more instruments, Euphrosyne is made to express a bold eagerness "to try the Experiment upon me."[108]

In *The Young Gentleman and Lady's Philosophy,* first issued in 1755, Martin earlier wondered why more women had not discovered happiness in the "experimental Spectacle of the Sciences." "Philosophical Entertainments," designed for the cabinet and classroom of the gentry and facilitated by the very optical instruments he manufactured, were touted as being "in the highest Degree entertaining and useful by the most easy and obvious Experiments."[109] Yet, in the end, the male mastery over fashionable apparatus mimicked the larger societal control over women who consumed fancy products without fully possessing them intellectually.[110]

In sum: portable "recreating" instruments (fig. 55) and an ever widening range of cheap illustrated books for children and their guardians continued to proliferate. There was no scientific subject that did not receive a generalized or specialized treatment. Reflecting the drive of shopkeepers, artisans, and tradesmen for self-education, useful recreations required work and similarly demanded interactive participation.[111] Even on the threshold of the nineteenth century, these gainful pleasures continued to promote the wisdom of the creator, as in Adam Walker's *System of Familiar Philosophy* (1799). This Lockean view of child development, expounded in lectures given to appreciative audiences in provincial towns, supposed that a child's amusing itself by trying to catch hold of everything in and out of its reach was a natural propensity. And this "instinct is the voice of God." The business of the teacher of experimental philosophy thus had nothing to do with "the bitterness of satire [a gibe at the atheistic *philosophes*], or enchantments of theater, neither thundering eloquence, nor sublime inspiration [stereotypical Jesuit tactics]." Rather, instruction was to be ruled by the "simplicity of reason" governing Emmanuel Barlow's accompanying plates. *On Light* (fig. 56) schematically rendered the sun as the material fountainhead of illumination in the universe, throwing off particles from the reflecting bodies touched by its rays and projecting them into infinite space.[112]

If Walker's compendium represented the afterlife of the seventeenth-century physico-theological tradition, Sir David Brewster's short and deliberately popular *Treatise on the Kaleidoscope* (1819) was heir to the baroque penchant for "creating and exhibiting beautiful forms." Like Kircher and Schott, whom he openly admired, the Scottish scientist was smitten by the beauty of formal experiments in patterned colors. Kaleidoscopic visions were the romantic culmination of a hermetic and ingenious *ars combinatoria*. This "system of endless changes," however, was now a "rational amusement" produced by "a general philosophical instrument."[113] Importantly, Brewster's optical apparatus also showed that the romantics were not opponents but appropriators of the Enlightenment.[114] Rational recreations succeeded in turning visual pleasures into moral philosophy and optical games into meditative icons.

55

Adam Walker

*Astronomy*

1799

from *A System of Familiar Philosophy*

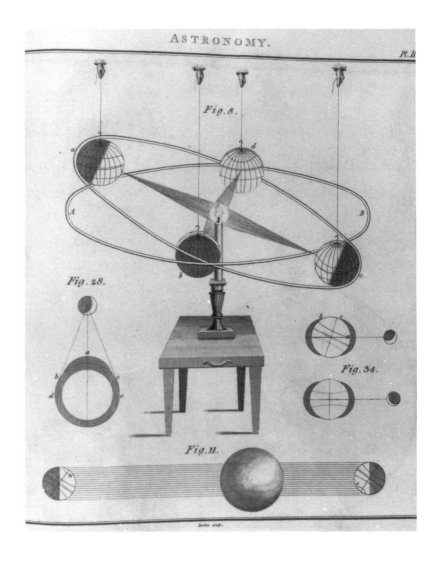

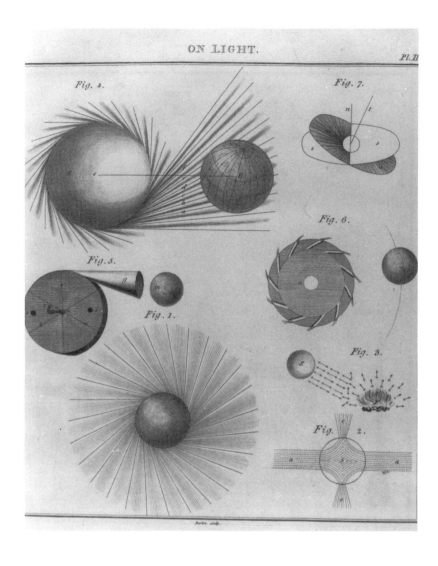

ON LIGHT.

*Homo* or, rather, *Puer ludens* thrived throughout the nineteenth century. The explosion of scientific toys was made possible by the notion first put forward in the seventeenth century that mathematics might be recreational. Instruments, or "ideas made brass," continued to draw enthusiastic audiences wishing to see the show.[115] Further, the educational and skeptical entertainments of the eighteenth century helped to fashion a flourishing market for small and simple science "playbooks." There was, however, an important difference between the two centuries. Scientific subject matter tended to become more cut and dried. While John Moffatt (1835) (fig. 57) and John Henry Pepper (1861) created attractive primers, their informational content reflected the increasing specialization of the sciences even at the most elementary level. Education was now conceived in terms of juxtaposed and competing domains. Metallurgy, geology, paleontology, chemistry, and microscopy were represented as isolated, autonomous, and sequentially structured disciplines in Landall's frontispiece.

This notion that certain skills had to be mastered before others could be acquired is the ancestor to the tracking and placement systems found in contemporary schools.[116] Gone is the eighteenth-century practice of cooperative learning across fluidly defined fields. Gone, too, is that earlier era's rewarding of individual expression and effort. Enlightened instructional practice, embodied in entertaining illustrated books, placed a high value on the imagination. The person was schooled throughout life; creative thought was not narcissistic but dynamically engaged with the environment. Mobile, and so responsive to change, it could be placed at the service of a common enterprise.[117] Self-realizing performances enacted during youth were to shape adult minds, releasing them from idle vacancy and despotic superstition. Liberated at last, the intellect might aspire to reforming the social sphere by reconceiving it.

57

John M. Moffatt

frontispiece to *The Book of Science*

1835

58

Johann Samuel Halle

*Projecting Apparitions on Smoke*

1784–1786

from *Magie*

### Systems of Imposture

Rational recreation was visual education. Enlightened entertainment allowed for the legitimate indulging of the eyes in nondelusory patterns and mind-building shapes. International in appeal, this popular form of instruction relied on novel and sensuous technology. Here, however, was the rub. Optically communicated information comprised the stock-in-trade of quackish hosts of prestidigitators (fig. 58), operators, schemers, empirics, entrepreneurs, and instrument makers.[1] Rational recreation therefore was also the deliberate counterpoint to fantastic or *irrational* recreation. On one hand, the competitive leisure industry pressured the informing philosophical illusionist to distinguish himself from the deluding conjuror. On the other hand, popular educators relied on the same battery of stunning newfangled devices to attract the consumer's gaze.

The hazardous and contested boundary between serious and spurious forms of culture mirrored the larger ambiguities of what it meant to be a professional or a "mechanic" during the eighteenth century. Charging money for services, deploying ostentatious equipment, but above all playing on visual appeal raised the old problem of the role of the eye in cognition. At issue was the ancient question of what constitutes bad teaching. Plato's ethical argument against the Sophists had accused them of using the corrupting spell of *rhetoric* to please and entertain the idle rich rather than the noble virtue of *philosophy* to train ethical citizens for the healthy polis.[2] The gratification of the base "common" senses was not to be confounded with high intellectual enjoyment.[3]

For eighteenth-century thinkers, the role of pleasure in the formation of an "educated sensibility" remained troublingly equivocal. This ancient problem was compounded for the moderns by the proliferation of useful and delightful instruments. Both the instructor and the mountebank manipulated gadgetry to visibilize an invisible realm. Matters were further muddied because this host of attractive devices was uncontrollably accumulated by the young, their parents, and an ever-widening middle-class public. The challenge for enlighteners, then, was to fill leisure hours pleasantly and productively while paradoxically relying on sophisticated fairground apparatus.

The Age of Reason was keenly aware of the complexity of child's play. Games, romps, and fantasies of all kinds, as we saw in Locke, possessed an ethical as well as an aesthetic quotient. They could be interpreted as rehearsals, either for good or ill, of future adult activity. The still unanswered question concerning such simulation was how much imitative behavior owed to inborn predisposition and how much to socialization. Further, mimicry was itself a problematic category. If play eased the child's entry into group life, this integration was accomplished through pretence. The illusionism of ritualized pastimes and their communalism left such performances open to the charge of feigning and herd cohesion. This accusation was especially potent when apparatus intervened to produce intense physical and sensory stimulation. Optical devices, in particular, rendered ambiguous the distinction between reality and play-acting, benevolent intention and malevolent maneuver.

Indeed the reputation of machines, techniques, and their legerdemain "operators" was and remains begrimed.[4] The ancient primacy of theory owed to the pervasive value bestowed on invisible, priceless thought over visible, costly experiment, handless hypothesis over sleight-of-hand manipulation. The Platonic fear of "sophistication," that is, the rhetorical adulteration of noble ideas by ignoble and even deceptive tinkering, was itself a topic for discussion within the "philosophical entertainments" enterprise. It is to this decidedly moral component of the talk about "juggling" mechanisms and ingenious tricks for sale that we must now turn. The clever technician as the crafty fabricator of nonintellectual goods became the target of the *philosophes'* attack on fraud. Most fundamentally, to enlighten meant unmasking charlatanism of every stripe by teaching the public its conning stratagems.

The hauteur of many twentieth-century intellectuals toward dexterous practice or skilled execution in the arts and sciences finds its roots in the early modern period's identification of the "méchanicien, peintre, opticien," in Etienne-Gaspard Robertson's words, with the "magicien, nécromancier" (fig. 59). Like many earlier and later illusionists, this self-professed inventor of phantasmagoria took pains to separate superstitious sorcery and prodigies from his own rational trials, supposedly presented "without charlatanism."[5] Fear of contaminating pristine or authentic experience reached epidemic proportions in the second half of the eighteenth century. A defensiveness persistently lurked just under the surface of countless *mémoires récréatifs*. Like Robertson, these authors felt constrained to argue on behalf of practice as necessarily preceding, explaining, and testing theoretical claims. Thus the "instructive spectacle

of phantasmagoria," unlike the *diableries* of della Porta, van Etten, and Kircher, was offered to the public as a philanthropic "experiment." Looking back from the vantage of the Restoration and the July Monarchy to the perilous period just following the French Revolution, Robertson argued that, while cosmopolitan cities might be rid of the imaginary terror of gulling specters, such apparitions still lingered tenaciously in the superstitious folk customs of Catholic provinces.[6]

By examining what the enlighteners perceived as constituting a libertine "system of imposture,"[7] I am proposing an alternative to a current construction of the history and philosophy of science.[8] Embedding the unstable compound of performative experiment and out-and-out chicanery into the wider cultural debate over technology and visualization shows the inextricability of theoretical concerns from practical procedures. As we saw in chapter 1, we need to restore the study of religion to the Age of Reason. Similarly, logical game playing of all sorts undercuts the opposition supposedly existing between public and private realms in the old regime.[9] What makes the intersecting duplicities so telling is that they occur precisely in those domains seemingly least susceptible to rational control. Indeed, as the permeability of rational recreations demonstrated, private life took shape in Jürgen Habermas's public sphere. I want to argue specifically that the preoccupation with hypocrisy and deceit found within the literature of instructive games permits us to witness the complex intersection of public institutions with private sites of sociability, and not just in France.

Historians of statistics have shown how eighteenth-century rationalism launched an attack against the feudal metrology of the *ancien régime*.[10] The Calvinist-born Rousseau's war against insincere appearances must be interpreted against the backdrop of the gaudy sharpers and crooks who swindled people in the absence of standardized weights and measures (fig. 60). The rococo painter Pater's center-stage quack, selling dubious potions deceptively bottled in a flask that is both thin and wide, embodied the irregularity common to fairground and market transactions alike.[11] Thus eighteenth-century mathematical recreations served a purpose rather different from the ingenious problems set by Schott, Schwenter, and Bonanni. Calculating, as Leibniz had earlier recognized, freed humanity from corrupting charlatans, enlightening them to think for themselves.

60

Jean-Baptiste-Joseph Pater

*The Fair at Bezons*

c. 1733

61

Francisco de Goya

*Nobody Knows Himself*

1799

Reliability becomes a value in all cultures where the quality and monetary value of goods is not stabilized.[12] Goya satirized such Frenchified card-sharps and double-dealing Spanish *petimetres*. Operating in a deceitful world without commonly accepted norms, they were able to dupe the ignorant populace through a harlequin masquerade (fig. 61).[13] *Los Caprichos* vividly recorded the absence of quantifying reason in that vast and universal system of imposture that fed the groundless fantasizing of gullible onlookers everywhere.

Not only superstition but high-minded intellectual pursuits could become suspect. Steven Shapin and Simon Schaffer investigated criticism leveled at the British experimental program during the seventeenth century. Hobbes equated such public performances, filled with phantasm and illusion, with a confederacy of deceivers. For the materialist freethinker, the ingenious artificers of this "engine philosophy" were no better than the gross papist idolators and quackish alchemists who obtained dominion over the rabble through hermetic conjurations (fig. 62).[14]

For the early moderns an analogy existed between the legerdemain of experimentalists in all fields and the maneuvers of the con man. Like gapers at a magic show, science spectators seemed to participate merely by watching. The potential for fraud lurked in any demonstration in which the performer created the illusion of eyewitnessing without inform-ing the beholder how the action was done. When the viewer felt irresist-ibly magnetized by the visible invisible, especially in the presence of unreadable optical effects, then the technologist was condemned as a cheap trickster and his product judged as sophisticated flimflam. The total experience added up to commodified wonder selling base stupefaction.[15]

Thus juggling, as consummate mastery in the techniques of double-deal-ing, opens a new window onto the ethics, values, and expectations of a past era. As with faking, we catch a glimpse of the fears, desires, and apparatus motivating and mesmerizing a growing consumer society.[16] Understanding the theory and practice of deviousness also provides in-sight into how certain aesthetic ideals and artistic genres became suspect: speed, dexterity, wit, sophistication, clever reproductions, *faux* surfaces, *trompe-l'oeil,* and other rare or fashionable illusions (fig. 63). At a primal level, then, *tours de force* aroused the anxiety that processes of manufacture had fallen into the hands of uneducated artisans. These crafty men ma-neuvered outside academic control and engaged in deceptive manipula-tions to separate potential buyers from their money.[17]

62

David Teniers the Younger

*The Alchemist*

1649

63

Jean-Etienne Liotard

*Still Life: Tea Set*

c. 1781–1783

94    TESTAMENT

cartes, en ferrant le paquet inférieur entre le pouce & le doigt du milieu de cette main. Voyez la *fig.* 21.

Dans cette pofition, le paquet fupérieur fe trouve ferré entre le petit doigt de la main gauche & les deux doigts annulaires, & du milieu de la même main.

3° En tenant toujours le paquet inférieur avec la main droite fans ferrer le paquet fupérieur avec cette main, tâchez de tirer ce dernier avec la main gauche pour le faire pafler par-deffous leftement & fans bruit. Vous trouverez de la difficulté en commençant ; mais une heure d'exercice par jour pendant une femaine vous donnera à cet égard la plus grande facilité. Remarquez qu'immédiatement après la

DE JÉRÔME SHARP.    95

coupe, les paquets peuvent & doivent avoir des pofitions différentes felon le befoin :

1° Ils peuvent être réunis & n'en faire qu'un, comme dans la *fig.* 22.

2° Ils peuvent être croifés & pofés de biais l'un fur l'autre, comme dans la *fig.* 23.

3° Ils peuvent être féparés, & un dans chaque main, comme dans la *fig.* 24.

4° Ils peuvent être féparés par l'index de la main droite, & fe trouver tous deux dans cette main, *fig.* 25.

Reginald Scot's *Discovery of Witchcraft* (1584) offers an early example of the extension of the manual "art of juggling" to include all choreographed duplicity.[18] Conjurors' rituals, like spiritual acts of revelation, involved the "nimble conveyance of the hand" and took mainly three forms: the hiding and conveying of balls; the alteration of coins; and the shuffling of cards (fig. 64). Scot, and subsequent writers on these wizards at cheating, emphasized the dangerous expertise of jugglers who could so deftly "abuse men's eyes and judgments." "Knacks" and facile manual exercises consequently became synonymous with irrational visual "delusions, or counterfeit actions."[19]

Samuel Rid's *Art of Juggling* (1614) provides a rare glimpse into the history behind the introduction of immoral "oriental" routines into England and the continent. "Certaine Egiptians," banished from their homeland, excelled in "quaint trickes and devises." Their foreignness, unintelligible dialect, and alien mannerisms made them at once remarkable and suspect. Apparently unknown before the era of Queen Mary, such Gypsies swiftly became the topic of conversation because of the ostentatious "strangeness of their attire and garments, together with their sleightes and legerde-

maines." Englishmen soon swelled their ranks, learning "craft and cozen-ing" to gyp country folk through palmistry and fortune-telling.[20]

The early nineteenth-century British artist Samuel Colman depicted their still common tricking routines in his *St. James' Fair* (fig. 65). Poor and ignorant young girls were especially vulnerable to being cozened of money, silver spoons, and apparel by roving bands of handsome peddlers, smooth-talking tinkers, and their charlatan descendants. Further, Edward Villiers Rippingille's (1788–1859) *Bristol Fair* unflatteringly represented the stereotypical exotic looks and false finery of modern carnivalesque itinerants (fig. 66). In his morally tinged painting of English rural pas-times, a disheveled Romany rogue acts as a sideshow barker enticing strolling spectators to idleness. Significantly, this romantic vagabond was shown luring British innocents to an optical box manipulated by an outlandishly clad and witchlike accomplice. Gypsy juggling thus embod-ied the criminal antithesis of authentic performance, the satanic inversion of bourgeois respectability.[21] As the gawdy and dangerous "other" of Empire, these nomadic men and women of color threatened the stability

66

Edward Villiers Rippingille

*Bristol Fair*

19th century

of conventional economics by their singular transactions.[22] Figures of fascination and repulsion, they were the dark precipitate society tried to repress, the grotesque disturbance to business as usual. Such migrants of ill repute—associated with idleness and diversion from gainful employment—highlighted the dark side of the leisure industry. Amusement thus seemed to require control.

Both the seductiveness and the venality of these dunning entertainers, originating in the mysterious "East" and subsisting on the margins of licit finance, was brilliantly analyzed in William Hazlitt's essay on "Indian Jugglers." Stimulated by the great London attraction of the summer of 1813, he reflected on the sensational performances cramming in eager crowds at Pall Mall. The romantic author distinguished praiseworthy trials of skill, manifested in the airborne calisthenics of flashing brass balls, from merely ingenious "craft or mystery." Extending his admiration to include the legendary dexterity of John Cavanaugh, the famous hand-fives player, Hazlitt extolled him as a master of "cleverness" and "knack." "Adroitness and off-hand readiness" characterized this mechanical "genius in trifles."[23] Hazlitt's essay on the art of the visible invisible transcended its nominal purpose of examining "a sport for children" to formulate, instead, a poetics of virtuosity. Thus he drew a crucial comparison between the verbal speed and agility needed by the hack writer to make a lowly pun, epigram, or extempore verse, and the juggler's sleight-of-hand. These facile literary skills were "like letting a glass fall sideways off a table, or else a trick like knowing the secret spring of a watch."[24]

Torn between respect and contempt, however, Hazlitt ambivalently concluded that keeping four balls up in the air ("which is what none of us could do to save our lives") was either a frivolous or a miraculous capability. Like George Morland's (1763–1804) rustic scene of *Dancing Dogs,* there was something inhumanly automatic and ridiculous in the unedifying demonstration of art for art's sake (fig. 67). Hazlitt simultaneously decried the tasteless and vulgar "mechanical deception" of the popular mountebank while sublimating such precision in execution into "a mathematical truth."[25]

Faultless hand to eye coordination was a capacity shared by the acrobatic charlatan and the sublimest artistic genius. Aesthetic judgments concerning "vulgar" or "refined" performances thus cannot be divorced, as Pierre Bourdieu has shown, from a theory of culture that embraces practical

67

George Morland

*The Dancing Dogs*

1789

68

Théodore Géricault

*The Bull Market*

1817

life.[26] Romanticism was fundamentally about the creation of such dizzyingly complex structures. In Constable's or Turner's landscapes, the virtuoso management of finely divided pigments and conflicting energies is analogous to unnatural feats of skill. Similarly, Géricault, Delacroix, and Courbet ostentatiously wrestled with the multiple parts of their figural compositions to create a harmonious balance out of mobile physical forces (fig. 68).[27]

Returning to the Enlightenment, we must determine what constituted the physiognomics of fraud in European consumer culture before 1800. The many faces of charlatanism comprised an advertisement for precisely those goods of which society did *not* approve. Here, again, instructive games expose an unmapped public and private psychology of anxiety linking aesthetics with economics. Playful "turns," that is, *lazzi* or conjuring tricks, shed light on physical and intellectual provinces that were believed vulnerable to adulteration. Popular shows, experimental science, surgery, medicine, the industrial and fine arts, ritualized religion, and the rites of politics all reeked of mercenary imposture and fraudulent replication. Such counterfeits destroyed the aura of individual objects and disturbingly communicated the flux of things. We must follow Rousseau's lead, then, concerning "unnatural" transactions that make a quantitative, but not a qualitative, difference. He exposed the fundamental alienation accompanying the introduction of money into the realm of nature. Civilization was that arbitrary state of equivalency in which anything might be exchanged and reexchanged. Significantly, the corruption of the *double-entendre* was revealed by the Genevan philosopher to Emile while they were watching a mountebank's magic duck approaching and fleeing spectators at a fair.[28]

Nonliterate media, the textless theater of music, dance, and circus, were always and everywhere the most international entertainment.[29] Beginning in the seventeenth century, an exotic *mélange* of English, French, Spanish, Italian, German, and Dutch artists—augmented by the odd Turk, Chinaman, and Gypsy—crisscrossed the continent. Jacques Lacombe's *Dictionnaire encyclopédique des amusemens* (1792) evoked this ambiguous cast of *banquistes* who lived at the expense, and on the edges of, the public they duped. He vividly seized, from the perspective of the often-burned eighteenth-century consumer, the dangerously ill-defined and unincorporable outsider character of these cheats. Some were quacks, selling unguents for wounds or rusty nails to cure toothache. Some were monster-mongers displaying, for a fee, a cow to which they superadded a third horn or a large young man dressed like a woman and billed as a "giant." Some were animal exhibitors who charged for parading apes from Ceylon and leopards from Africa. But, to use the charlatan's slang, most just "had some dodge [*truc*] to do in the blokes [*pour roustir les gonzes*], that is, a hoax to trap honest people, & sometimes to pay their debts in counterfeit money; in

this kingdom, as in many others, there are good & wicked subjects, victims & chiefs [*coryphées*]."[30]

It is not coincidental that John Newbery's "Tom Telescope," and other rational recreationists as well, either denounced cards or reconceived them as instructive games. Sleight-of-hand was synonymous with the *banquiste*'s card tricks beguiling the lower classes. An anonymous eighteenth-century conjuror's "flick" book (fig. 69) captures this pseudoreligious atmosphere of inexplicable appearance and disappearance, the visible invisible of textless sleight-of-hand. Needless to say, effortless *sprezzatura* was an old scam. Reginald Scot described how the blank pages of a book could be painted with identical pictures. The edges, appropriately notched, permitted the conjuror's fingers to quickly locate the desired spot after the volume was rotated.[31] These acrobatic *tours,* literally "turns," required copying facsimiles of the same figures, which were reoriented and systematically distributed at seven-page intervals. The producer of the Newberry Library's rare exemplar heightened the dissimulation by representing the suits in *trompe-l'oeil.* Thus the unwary spectator thought he saw an actual knave of spades overlaid by a four of diamonds, or a king of clubs covered by an ace, flash before his eyes.

In addition to cards, flowers, and harlequins, the flick or flip book also contained pungent social satires. Roistering clergy (*l'abbé Tise, l'abbé Quillé, le père Turbateur, le père clus*), aggressive soldiers, and misogynistic views of gluttonous and stinking nuns (*la mère goule, la mère d'aillon*) reveal the folkloric stereotypes shaping the view from below. This Bakhtinian heterogeneous assortment of colorful patterns and grotesque drolls embodied a serio-farcical "carnival" sense of the world.[32] The oppositional logic of card tricks and card playing, conducted in street squares or dingy rooms, turned the seemingly permanent and lawful institutions of Western society monstrously upside down. This parodic nature of popular entertainments in general elicited from the Enlightenment establishment an analogously improbable caricature relying on anomalous, unexpected, and absurd situations. High-culture critiques of low life—such as those penned by Lacombe or Duchesne—demonstrated that vulgar games performed in the border zone of reality were perceived as subversive, an ethical affront to established authority.

Cards, metonymically standing for gambling, give evidence of this far-ranging eighteenth-century preoccupation with problems of practical morality.[33] *Le jeu* was the antithesis of lawful play. Like the stockjobbing leading to the South Sea Bubble crash of 1720, paper credit and unregulated greed dogged the addicted player.[34] Henri-Gabriel Duchesne's *Dictionnaire de l'industrie* (1776) lamented that this distraction of the upper crust quickly degenerated into an uncontrollable passion.[35] Pater's amorous card-playing couple, seated in the lower right of an amusement-filled *fête galante,* enact this ritual of erotic and financial deflation. The aristocratic sport of love is shown to be no more enduring than the perilous game of chance (fig. 70).

From William Hogarth's moral progresses to the Nuremberg painter-engraver Michael Rentz's enlightened emblems,[36] the consumer's greed for windfall profit was judged as illicit recreation leading to certain ruin (fig. 71). The *philosophes,* in particular, connected gambling to the entire system of entertaining imposture governing the unenlightened segments of the *ancien régime.* Jacques Lacombe, inveterate compiler, bookseller, and art critic, scoured the contemporary literature for trenchant satires of this luckless sport. In his *Dictionnaire des amusemens,* he recounted Henri Decremps's tale of an *académie de jeu.* Composed of former classmates at a Jesuit *collège,* this institution was neither "a company of *savans,* nor a literary society, but simply an academy of gambling made up of sharpers [*aigrefins*] of all kinds who themselves were alternately cheats & rascals."[37]

70

Jean-Baptiste-Joseph Pater

*Fête galante*

18th century

71

Franz Anton, Graf von Sporck

*The Gamblers*

1767

from *Todtentanz*

Lacombe fleshed out this knavish clan with other *faiseurs de tours,* such as the fictitious "Sieur Pilferer," improbably discovered lounging on a street corner at the Cape of Good Hope! A patent invention, this native of Bohemia, doctor of pyrotechnics, and professor of chiromancy was given the equally derogatory English pseudonym of "Crook-Finger'd Jack" in the American colonies. The fabled swindler purportedly visited all the European academies to perfect himself in the popular sciences: algebra, mineralogy, trigonometry, hydrodynamics, and astronomy. Pilferer also voyaged throughout the civilized and half the savage world to familiarize himself with occult, hermetic, and transcendental philosophies. Lacombe informed the reader that his anti-hero was an adept in the cabbala, alchemy, necromancy, divination, the interpretation of dreams, and super-stition. Nor did he neglect such humbler tasks as curing toothache. Unlike vulgar empirics, however, this metaphysical quack did not yank out the painful part. Rather, following Kircher's principles of palingenesis, he cut off the sufferer's head![38]

"Very subtle" card tricks (see fig. 64) were part of that larger baroque oral-visual culture of hocus-pocus derided by the Enlightenment (fig. 72). The extemporizer (*improvisateur*), juggler (*jongleur*), and turner of tricks (*faiseur de tours*) all followed the same mesmerizing method. Such charm-ers' "deceptive arts" embodied Jesuit aesthetics as judged from the vantage of neoclassical rationalist criticism. Lacombe's mocking rules of the con-ning game were themselves a parody of stereotypical Counter-Reformation mimetic strategies. Never tell the audience in advance what piece of legerdemain will be performed. Have more than one way of creating the same illusion. Never repeat the identical maneuver at the spectators' request for fear of discovery. Only read books to ascertain whether your invention was already known to others. But above all, virtuosity and *fa presto* execution were emphasized. Lacombe claimed that the success of *tours d'adresse* involving conjuring, counting, body tricks, the collusion of a confederate, depended upon novelty, variety, and rapid succession. The viewer was constantly destabilized, believing she or he beheld new com-positions that were actually old combinations now artfully altered.[39]

The preface to Henry Breslaw's *Magical Companion* (1784) indicated that magic, not just in France but also in England, was looked upon as fascinating the "lower classes," aiming "to make the childish laugh with the tricks of Hocus Pocus and Legerdemain." The author, however, showed his awareness of the rational recreation genre in dedicating this printed

72

Philip Astley

*Magician "Nailing" Cards*

1785

from *Natural Magic*

grab bag of singularities to that inveterate collector, Sir Ashton Lever. Like the Ashmolean Museum's "innocent diversions," Breslaw's compendium was not intended to encourage vice and idleness. "Many useful inventions in Mathematics have owed their rise to some of these fanciful exhibitions." Indeed, his spectacle of popular tricks was filled with the antimiracle rhetoric of eighteenth-century Protestantism. The book's professed goal was to expose a self-serving "Black Art," working "upon the senses." Hence it protected the consumer from mountebanks engaged in dubious farces no different from papist "conjuring" (see fig. 6). "Various are the tricks, and fancies to amuse and surprises made use of by the nimble-fingered gentry, who exhibit for a livelihood, and gain a comfortable subsistence from their gaping audiences."[40]

The crude frontispiece to Philip Astley's *Natural Magic* (1785) captured this complex, and by no means exclusively lower-class, culture of staring, summed up in the motto "Inspicia et Judica." Both book and plate, pirated from Henri Decremps, illustrate the intricacies of the reformist attitude toward "exhibitors." Breslaw and Decremps had registered a sensitivity to the hierarchies structuring the ambiguous profession of the conjuror. These education-minded authors wished to distinguish base "mechanics," who studied "what is trifling more than what is useful," from recreating instructors, who undeceive eyewitnesses by exposing the mechanism behind the spectacle.[41]

On the contrary, Astley (1742–1814), the founder and proprietor of the renowned Circus of England, was more protective of his funambulists, clowns, and tumblers. In the competitive leisure industry of Hanoverian London, he understandably felt that magicians should not reveal the scenarios or "plots" whereby they earned a living. Familiarity with fringe theatrical entertainment, stocked by contortionists, ropedancers, and mimes, made this impresario wary of revealing the secrets of closely guarded routines.[42] Easily sated by masquerades, dumb pageants, waxworks, and outlandish monsters, fickle audiences hankered after originality. Successful marketing precluded "enlightening the Public so far as to diminish its Pleasures, by giving a Digest of Entertainment in which the Imposture of the Actor, and the credulity of the Spectator were equally necessary, and of which the Charm entirely consists."[43]

Henri Decremps's (1746–1826) numerous publications on "white magic" shed light on this new and fluid group of "cunning men" and "pretty puzzlers" trying to make a fortune in the late eighteenth-century capitals of Europe. Trained in Catholic theology at Toulouse, he abandoned orthodoxy under the influence of the occult experimentation unleashed by Mesmer and Cagliostro. The French hack writer emigrated to London during the Revolution but was expelled for his republican ideas. Like Robertson's and Salverte's later memoirs, Decremps's prolific books and pamphlets on "la science sansculotisée" claimed to combat superstition by exposing ocular fraud.

*La Magie blanche dévoilée* (1788), unillustrated except for the frontispiece, should be considered a manual of visual education, the imaginative complement to the rational recreation. In these nonmystifying illusions, the senses were to be momentarily confused, provoking the conversation of the cultivated members of the audience while piquing the curiosity of the uncultivated. This small and cheap compendium, also published in English and Italian editions, ostensibly responded to the desire of intelligent readers to understand the maneuvers deployed to enchant them.[44] As noted, Astley's frontispiece schematically reproduced Decremps's enlightened prestidigitator, but with the important addition of bored, beguiled, and skeptical onlookers. Standing between two popular Parisian automata, the *Wise Turk* and the *Singing Bird* stoppering a bottle, the Chevalier Pinetti de Merci is caught in midperformance. Having thrown a deck of cards into the air, the internationally renowned illusionist proceeds to "nail" them to the wall with pistol shots. Significantly, many of the trick cabinet pieces and the palingenetic demonstrations presented by the conjuror and recorded by Decremps relied not on passive voyeurism, but on audience participation.

No illusion was more cosmopolitan in popularity than the resurrection of a dead bird (fig. 73). Using the broken remnants of an earlier exhibition (three eggs "danced" on the end of a cane), the magician glued together two of the larger halves. When these were cracked again during the act, a live canary emerged. In Greuzian *sensible* fashion (fig. 74), "a Lady of the Company is required to take it in her hands and, shortly after, the bird dies." The prestidigitator, as seen on the title page of the fifth volume of Johann Samuel Halle's *Fortgesetzte Magie,* then placed it under a bell jar containing "dephlogisticated air."[45] After a few moments, the victim revived and flew away. Decremps unmasked this apparent rebirth by telling how the pricking of pinholes in advance allowed the canary to breathe while in the shell. When the bird was handed over to the young

**Fortgeſetzte Magie,**
oder, die
**Zauberkräfte der Natur,**
ſo auf den Nutzen und die Beluſtigung
angewandt worden,

von
**Johann Samuel Halle,**
Profeſſor.

Mit 5 Kupfertafeln.

_pag. 323._

_Halle f._

Fünfter Band.

Berlin, 1793.
In der Pauliſchen Buchhandlung.

73

Johann Samuel Halle

_The Art of Conjuring with a Bird_

1793

from _Fortgesetzte Magie_

74

Jean-Baptiste Greuze

*Girl with Doves*

c. 1800

woman, however, it was killed "by the pressure of thumb and forefinger" on its throat. Since the bell jar covered a trap in the table top, an undetected accomplice could substitute a living animal.[46]

The modern reader is struck by the frequency of violent hoaxes found in such "entertaining" accounts. Typically, a cruel death is meted out which is then followed by an amazing recovery bypassing any religious intervention. Doves are decapitated, portraits of brothers are gouged, arms and stomach of the magician are stabbed. This murderous physicality was matched by an equally Greuzian exquisite sensuality (fig. 75). In the illusion called "the Magical Nosegay," budding flowers and shriveled fruit wondrously expanded and swelled like aerostatic balloons. Plants calcined to ash and then heated in a hermetically sealed flask miraculously assumed their original form. Concealed bellows produced this sexualized "palingenesia," or eroticized regeneration, creating "new objects for the sight of the spectators."[47]

These sly rococo operations—connected to the deceitful maneuvers of the charlatan—conspicuously differentiate Greuze's mobile *tours d'adresse* from Wright of Derby's frozen gestures and deliberately stilled hands (fig. 76). David Solkin has argued that James Ferguson is "producing society" by teaching his audience about the character of their own world.[48] What is distinctive about this scientific performance, however, is that it is represented by the British painter as separating itself both from spontaneous gestures and from the rhetorical motions of "showmen." These by no means extemporaneous performances were part of a systematic gestural language designed in advance as deliberate elements in a choreographed presentation.[49] Not only virtuosity, then, but specifically French codes of masked deceitfulness and delicate prurience were excluded from Wright of Derby's picture. Through a powerful visual argument the laboratory, unlike the duplicitous theater, became the site of "objective" experience and unfeignable sensibility.[50]

No amount of rehearsed legerdemain, Wright of Derby seems to be suggesting, could elicit the grave and moving behavior of the mixed audience observing the *Experiment with the Air Pump*. While scholars have doubted whether the science demonstrator used a living bird,[51] I believe the power of the painting consisted in the dialogical differences setting it apart from the crafty but only apparent rescues of the popular magic show. The nuanced dramatization of viewer empathy depended precisely

75

Jean-Baptiste Greuze

*Mademoiselle Madeleine Barberie de Courteille*

1759

76

Joseph Wright of Derby

*Experiment with the Air Pump*

1768

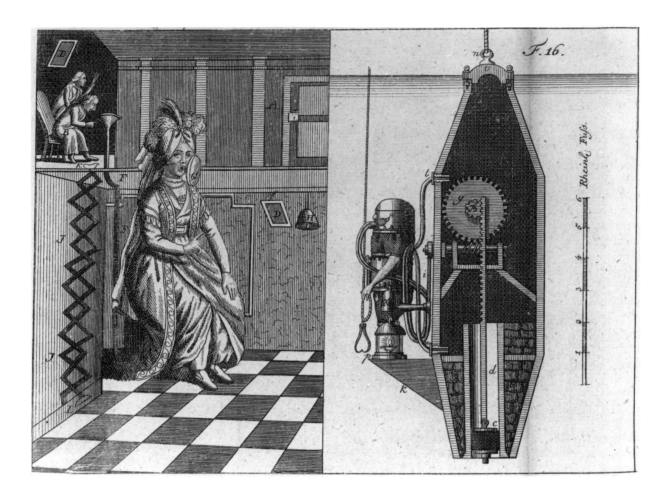

upon the real and shocking killing of the white cockatoo for whom there would be certain death if the stopcock was turned. Further, the carious human skull, sunk within an internally glowing glass beaker, also distinguished itself from the phosphorescent tricks of the conjuror. Wright of Derby tried to persuade the viewer visually that experimental science demonstrated ineluctable mortality and inevitable pathology, not the guileful illusions forming the mountebank's stock-in-trade.[52] Unlike Astley's or Halle's distracted gapers, pensive onlookers crowded around the empirical stage, collapsed into a table top without false bottom. This straightforward platform signified authenticity by forcing the observer to confront close up an inescapably instructive entertainment.

Like Decremps, Halle, a professor at the Royal Prussian Military Academy, argued that his seventeen-volume *Fortgesetzte Magie* (1788–1802) was an Enlightenment project. "Aufklärung" characterized his mission to establish boundaries between "deceptive enthusiasms" and the physical sciences. This illustrated compendium thus extracted "useful recreations" (*nützliche Unterhaltungen*), culled from the Swedish Academy of Science's forty-one volumes of Transactions.[53] In addition to Pinetti's playful "Kunststücke" involving self-firing pistols, evaporating inks, dancing jewels, and flowers reviving from their ashes,[54] practical automation found its way into the series. The *Brahmin, or Speaking Oracle,* trumpeting "supernatural lies" by means of tubes concealed in wall ducts and men hidden on a balcony, was pointedly contrasted to the handless and apparently guileless automation of a *Diving Bell* (fig. 77).[55] This moral view of automated apparatus as "sincere" machinery suggested that instruments existed capable of performing without the intervention of human adroitness or knack. Self-operation thus formed the counterpart to such popular sham mechanisms as talking parrots, aria-singing canaries, flying pigeons, eating ducks, and bell-striking Turks.[56]

### A Culture of Operators

To *operate* could signify "a methodical application of the hand or instruments upon the human body" performed by knowledgeable surgeons.[57] More generally, *operation* had the aesthetic connotation of "realizing" something for the eyes, as in well-managed light effects.[58] But the foreign impostors, exotic charlatans, and vernacular quacks—congregating at fairs, on turnpikes, and around pissing posts—gave execution a bad name. The accusation of "juggling" threatened such professions as medicine, experimental science, and the applied and fine arts that depended specifically on manual skill.

It has not been remarked that surgical reformers like William Cheselden, in the monumental *Osteographia* (1733), required his artists to represent bones without showing the forming hand behind the image (fig. 78). "Objectivity," or the honest conduct of the practitioner, was thus synonymous with the absence of any visible sign of manufacture.[59] The rise of objectivity as a scientific ideal in the early modern period was facilitated by the development of measuring and distancing apparatuses.[60] These truly "automatic" devices seemed to preclude shady handling and phony gadgetry (fig. 79).[61] In sum: The dialectical tension between "well-performed" or impalpable demonstrations and staged ruses—involving conspicuous stroking, touching, and finger pointing—must be embedded

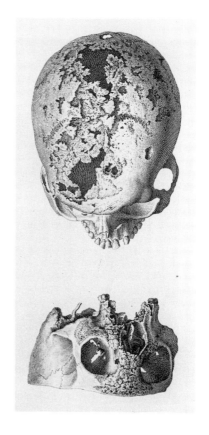

78

William Cheselden

*Skull of a Woman*

*Who Died of Venereal Disease*

1733

from *Osteographia*

# OSTEOGRAPHIA,

## OR THE

# ANATOMY

## OF THE

# BONES.

BY WILLIAM CHESELDEN

SURGEON TO HER MAJESTY;

F. R. S.

SURGEON TO Sᵗ THOMAS'S HOSPITAL,

AND MEMBER OF THE ROYAL ACADEMY OF SURGERY AT PARIS.

LONDON MDCCXXXIII.

80

Johann Wilhelm Meil

*The Engraver*

1765

from *Spectakulum Naturae & Artium*

within the broader culture of fashionable fraud. Animal magnetism
summed up this "modern theurgy" run by professional operators who
juggled with the invisible as if it were a physical agent.[62]

Analogously, the mechanical production of multiple impressions, replicas,
and variants could look like visual quackery to the creator of painted
originals (fig. 80). Indeed, the custom of treating prints with Venetian
terebinth in order to make transparencies was identical to a "quackish"
recipe against syphilis![63] In an age of forgeries, Hume considered it
presumptuous to conclude that ink marks on paper constituted testimo-
nial evidence of any sort.[64]

Gilles Deleuze has argued that the simulacrum, in a "reverse Platonism,"
denotes the triumph of the pretender. The phantasmatic effect, fabricated
by intervening machinery, breaks down the division between pure essence
and impure appearance, model and imitation, to vindicate the inauthentic
image without proper parentage (fig. 81).[65] Sophistic operations negatively duplicate the charlatan's wholly *aesthetic* claims. They playfully
present false visual evidence with no concern for ethics. Baroque optical
displays, such as Kaspar Schott's *Anamorphosis,* were similarly contrived to
deviate from the dialectic between regular and irregular shapes, thus

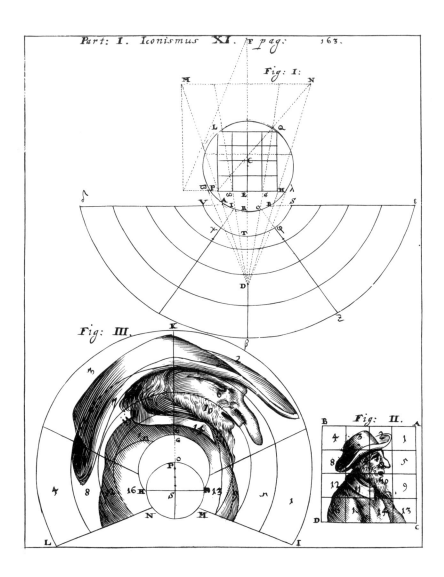

highlighting their nonhierarchical coexistence. Mutually incompatible verisimilar illusion and dissembling delusion were forced to become compatible, to refer ambiguously to one another as equals. Schott's grotesquely skewed punchinello can be looked upon as a symbolic metamorphosis of a "natural" likeness. Conversely, the portrait of a bearded and dignified old man can also seem anomalous, the serious variant of a ludicrous figure. By a crafty sleight-of-hand, cylindrical and conical lenses were made to rectify deformed images or to scramble resemblances.[66]

Like the visible invisible of the conjuror, contrived reflections simply appeared and disappeared, wrenched in and out of shape by the behind-the-scene motions of the operator. Speaking of that supremely romantic device, the "Kaleidoscope as an instrument of amusement," not of science, Sir David Brewster warned against multiplying objects by mirroring. Citing della Porta's, Kircher's, and Schott's glassy show boxes, he remarked that accidental "colour independent of [essential] form is incapable of yielding a continuous pleasure." Shifting "masses of rich and harmonious tints, following one another in succession, or combined according to certain laws, would no doubt give satisfaction to a person who had not been familiar with the contemplation of colours; but this satisfaction would not be permanent, and he would cease to admire them as soon as they ceased to be new."[67]

This Platonic construction of reality, inherited by the romantics from Enlightenment philosophy, was predicated on an oppositional dichotomy. If the "original" or first model is "real" or "true," then the subsequent copy perforce must be unreal or false. The aesthetic implications of such a dualistic logic can best be analyzed by looking at the problem of duplicates.[68] Material goods, including replicated images, were open— both literally and ontologically—to the accusation of being ungenuine. By definition, manufacturing was artificial, an unnatural tinkering with nature in order to produce semblances.[69] Hence the suspicion of fraud was inseparable from the activity of copying (fig. 82). Like all hand- or machine-made likenesses, the surface of valuable artifacts might be mimicked, not just for instruction or edification but for money. Deft simulations supposedly lacked the invisible depth and the priceless spiritual immateriality of the authentic work they sought to reproduce.[70]

Savery de Bruslon's (1657–1716) posthumously published *Dictionnaire de Commerce* (1742) registered French officialdom's anxiety concerning the prevalence of "hidden swindling." His long article on "Fraud" ranged from the perils of contraband, to cheating the king by failing to pay the hated *gabelle,* to chicanery in business affairs when debtors dodged creditors. Connivance reigned in the cloth industry, for example, where silk or wool might be woven containing an insufficient number of threads. Not surprisingly, this Inspecteur-Général des Manufactures under Louvois witnessed an enormous volume of diverse goods passing through the Paris customs house. The subversive economics of *trompe-l'oeil* adulteration colored Savery's cross references: *fraudeur* (a person who gains his livelihood by bilking), *frauduleusement* (to do so in a "scandalous" manner), *frauduleux/euse* (cases in which things were deliberately made to deceive).[71]

82

Johann Heiss

*The Sculpture Academy*

17th century

Such ruses were fed by the restless taste for *nouveauté*. The ever-changing category of the new, or of that which had not yet appeared, ranked among the inventions of commerce. Novel fashions in shawls, ribbons, and hair styles were advertised in posters and displayed daily by shopkeepers wishing to tempt the luxury trade.[72] Goods that appeared or disappeared, depending on the whim of sellers, were a form of reverse pickpocketing. Instead of adroitly removing watches, cutting off buttons, or lifting clasps, the merchant subtly conjured the desire for frivolous objects while spiriting them into the customer's hands.[73]

Technical shortcuts, especially, were criticized for evading honest labor bought and paid for. Legerdemain practice had long been a hallmark of baroque cunning. Casting leaves from nature for Bernini's great baldachino in St. Peter's, or creating intricately wrought Dutch and Flemish *pronkstilleven,* teetered uncomfortably between virtuosity and counterfeiting.[74] Neoplatonically minded critics, in particular, found that ingenious ideas and dazzling skill could seem indistinguishable from trickery. For the eighteenth century, however, it was the print world that constituted a monstrous danger to "the exquisite and extraordinary" *chef d'oeuvre.*[75] Multiple impressions corrupted singular masterpieces by fetishizing them. Serial production turned rare originals into "objets de caprice." Like the fetish, the facsimile duped by promising marvels while delivering only the effects of divinity.[76]

Any reproduction of a visual work of art creates a representation that fails to capture some properties of the original (fig. 83). Like today's digital image storage discs, the most common eighteenth-century methods of reproducing art shared the property of being passive with respect to the content of the image. Thus Elizabeth Vigée-Lebrun's pupil, Marie-Victoire Lemoine, "stores" her teacher's composition in a copy without personal enhancement. Later in her career she will "retrieve" this encoded drawing both in memory and practice, to insert it, now reconstructed, into an altered imitation or design of her own invention.[77]

Academic training, as it occurred in the private studio and the official school (fig. 84), was based on the assumption that because the student internalized figures, he (rarely she) could subsequently convert that knowledge into an authentic work of his own.[78] The avoidance of forgery was achieved by increasing the distance between student and master.[79] Jean-Auguste-Dominique Ingres's *Raphael and the Fornarina* demonstrates

83

Marie-Victoire Lemoine

*Mme. Vigée-Lebrun and Her Pupil, Mlle. Lemoine*

18th century

84

Johann Heiss

*The Painting Academy*

17th century

85

Jean-Auguste-Dominque Ingres

*Raphael and the Fornarina*

1811–1812

the complexity of this alienating procedure (fig. 85). Reconstructing the semblance of the young Raphael from copies he had made of his self-portraits, the nineteenth-century French artist shows the Roman painter clasping his mistress while gazing upon her sketch. The viewer of Ingres's composition can see, however, that the likeness between "live" model and drawing on the canvas is only approximate. Raphael, Ingres seems to be suggesting, transformed the flesh-and-blood beautiful baker's daughter into a new construction. Yet, dimly visible and leaning against the back wall, we glimpse a portion of the *Madonna della Sedia* whose wrapped head, seated pose, and V-shaped cradling of the infant Jesus Ingres borrowed. The turbaned Fornarina thus is neither her "pure" historical self nor the original of the representation on the easel. She is a fictional variant of Raphael's madonna, a simulacrum who already exists within the artificial realm of an earlier picture.[80]

The hermetic world of fine-art copying, captured by Lemoine and Ingres, was based on the aristocratic emulation and transmission of select exemplars passed from gifted masters to their worthy descendants. It was the means to an end (invention), not the end itself. Industrial developments, however, increasingly attacked the unique products of painter, sculptor, and architect.[81] *Trompe-l'oeil,* fake materials, the cheap serial counterfeiting of goods, even amateur art instruction seemed a simian prostitution of the perfection and difficulty of art for which the mythic Raphael stood.[82] Mechanical processes, like the charlatan's *tours de passe-passe,* corrupted the high aesthetic of *bella mano* or the inimitable refinements of a particular hand. Baboonlike, the copyist facilely recorded the impressions conveyed by a machine (fig. 86).[83] Contrary to Ingres's Raphael, who individualized and sensualized the act of holding a brush, the rote transcriber's disembodied fingers merely repeated the virtual image of an external object projected by means of a mirror or prism on the plane surface of a camera lucida. The hostility toward exhibitionism and manifest feigning, evinced by the Enlightenment against all those who made a pretense, revealed its fundamentally antitheatrical bias.[84]

Lacombe satirized this garish mercenary in the whimsical character of the adroit "Sieur Malpigiani." Advertising like a quack, the fictitious "Mr. He Who Paints Ill" claimed he could teach Parisians how to draw in three lessons for the modest sum of one louis. This *fa presto* Italian virtuoso apparently was so versed in the principles of art that he could trace, "in an instant," on the sand with his foot or a stick the portrait of anyone.

86

Martin Frobenius Ledermüller

*Solar Microscope*

*Used with a Camera Lucida*

1768

from *Amusemens microscopiques*

Lacombe's witty description negatively evoked the notorious rococo *sprez-zatura* of his contemporary Jean-Honoré Fragonard, torn between the virtue of craft and the virtuosity of speed.[85] This *peintre habile,* playing on the instability of sight, resembled those *joueurs de gobelets* plying their trade at outdoor fairs in front of "eighteen hundred" spectators. Such thim-bleriggers and fleecers (*escamoteurs*) ran a scam for quickly copying a likeness. Their ruse involved "spoiling a very good print to make a very bad picture." Aptly, Lacombe saw the commerce in cheap portrait engrav-ings as enabling Malpigiani's sleight-of-hand.

Fragonard's palimpsestic records of personalities, reputedly painted in just one hour, artfully appropriated the maneuvers of the carnival trickster. Like the dissembling huckster's, his mimicry led not to the proper reve-lation of the "other" but only to the ostentatious display of despotic skill. The rococo artist's pictorial pranks deliberately challenged a fundamental assumption undergirding the practice of portraiture. He disturbingly denied that the fabricated copy contained a deliberate allusion to a human original.[86]

Fragonard's impersonations were hardly face painting but a parody, I believe, of the mountebank's deauthenticating artistic practice. No doubt these equivocal doubles were intended for the sophisticated delectation of his *fermiers-généraux* patrons, capable of appreciating the jest. One example must stand for this brilliant series, a long meditation on the problems of constructing an identity. Conspicuously applied brush strokes "make up" the artificial person staring out of the Art Institute of Chicago's picture (fig. 87). The ill-fitting paper-doll Spanish costume, implying easy re-moval, and the clearly borrowed features overtly proclaim the work as fiction. All elements of the composition pretend to be what they patently are not, i.e., rapidly painted and referring to a specific character, perhaps Don Quixote.[87]

The exotic or fancy-dress quality of Fragonard's "sitter" would have an-nounced to eighteenth-century viewers that his identity preexisted this most recent Spanish expression. In fact, the stereotypical face and torso might well have been derived by overlaying a seventeenth-century tem-plate. Possibly a costume piece served as pattern. Thus neither this paint-ing nor the thirteen other so-called *portraits de fantaisie* can be interpreted as documentary evidence of actual men and women. Rather, they were invented lives.[88] As imaginary visual biographies, such comic simulations masqueraded as copies of well-known contemporary or historical types.[89]

87

Jean-Honoré Fragonard

*Portrait of a Man*

c. 1767–1770

Joseph-Aignan Sigaud de La Fond

*Electrical Experiments*

and the *"Magic Portrait"*

1775

from *Description et usage*

*d'un cabinet de physique*

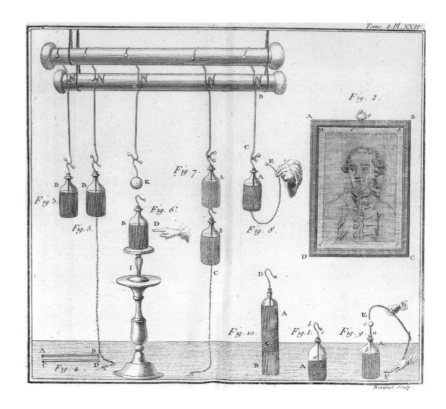

It seems plausible that Fragonard was alluding to a common deceit. Malpigiani, according to Lacombe, doctored an engraving by soaking it in cold water overnight. To produce a "tableau magique" (fig. 88), he placed it face down on a sheet of Bohemian glass, wiped it with Venetian terebinth, and lightly rubbed the back to leave a faint imprint behind. This outline, invisible to the audience, could be overlaid on fresh paper and filled in on the spot with the desired features of the unsuspecting mark.[90] Goya, in the domain of "original" etchings and not "reproductive" engravings, also drew attention to this legerdemain capacity of all impressions to appear and disappear (fig. 89). In *Be Quick, They Are Waking Up,* from the *Caprichos,* the sweeping monk literally duplicates the etcher's repeated gesture of first inking and then wiping the plate. Black and white media, especially, were capable of juggling with cinematic light and shade.[91] The elusive sphere of replication was thus fundamentally allied with conjuring, with the transitory opticality of the variety show and the vanishing act.

Reproduction, then, could be made synonymous with ignorance, with an "illiterate" technology that evaded proper training.[92] The eighteenth cen-

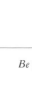

*Despacha, que dispiertan.*

90

Nicolas Bion

*Compasses*

1758

from *The Construction and Principal Uses*

*of Mathematical Instruments*

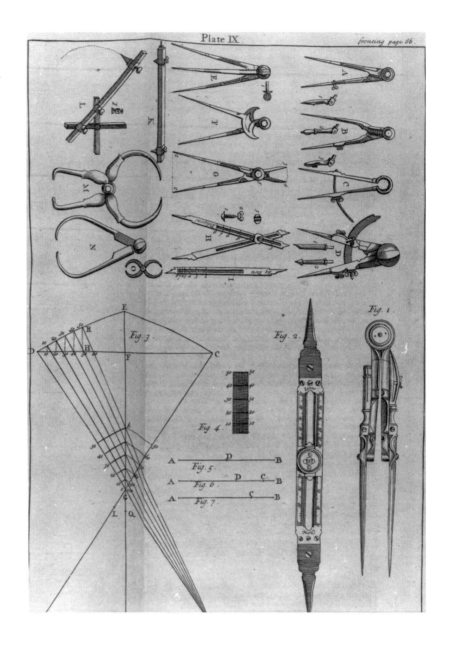

CABINET D'UN PEINTRE

*Dédié à Madame Marie Henriette Ayrer, veuve de feu Mr. G. Chodowiecki.*
*Par son très humble et très Obeissant Serviteur et fils*
*Daniel Chodowiecki.*

tury was torn by polemics concerning artistic ownership. Debates and even lawsuits raged over who invented the "secrets" of pastel fixatives, color printing, and time-saving copying devices.[93] Advances in artifice were sometimes difficult to separate from swindling. Machine-made portraits, silhouette-taking contraptions, and compasses for reducing the dimensions of a painting for transfer to another medium (fig. 90) endangered the livelihood of professional artists (fig. 91).[94] On one hand, Daniel Chodowiecki (1726–1801), the famous Berlin engraver, could hope to increase his output profitably by relying on multiplying processes.[95] On the other hand, the *automatic* pentograph and physionotrace, guaranteeing "absolutely faithful likenesses in even the smallest detail," could dispense entirely with his costly and painfully acquired expertise in drawing.[96]

91

Daniel Chodowiecki

*The Artist's Studio*

1771

EMM. JOS. SIEYES.                    ÆTATIS SUÆ. 69.

92

Jacques-Louis David

*Portrait of Emmanuel Joseph Sièyes*

1817

The rise of unskilled labor in the marketplace, such as *banquistes* at the *foire,* could lead to the seizure of jobs. Erudite, academically trained artists were liable to be usurped by Malpigianis armed with the latest equipment. This implicit threat was the subtext of Lacombe's satire on the "so-called draftsman" (*prétendu dessinateur*) who was capable of recording a face "in one minute," or quickly throwing off "quelques têtes de fantaisie." Surely the specter of the automaton-draftsman (*automate dessinateur*)[97] lies behind the deliberate probity of much neoclassical art and theory, proclaiming "solid and valuable things." Physical intensity signified moral rectitude, and thus privileged the invisible realm of the *je ne sais quoi.* As in Jacques-Louis David's somber *Sièyes,* the unfeignable grasp of deep character was at the antipodes from surface impressions mindlessly captured by pleasing techniques and their monkey mechanics (fig. 92).[98]

### *"The Prince of Charlatans"*

The automaton was the machine as virtuoso (fig. 93). Its anthropomorphic physiology of concealed pipes and pulsating liquids established a metaphorical connection between animated artifice and live audience. From the perspective of the entranced spectator, the magical work appeared to be composing itself from out of its own material substance. Analogously, the sly conjuror not only surrounded himself with this ingenious clockwork but was similarly self-motivated and autodirected. This *méchanicien* conceived of himself as an artifact, a dazzling producer of incantatory effects that magically captivated and controlled others by their showiness.[99]

Juggling, then, transcended wise Turks, bird organs (*serinettes*), and mechanical orchestras, to signify a disingenuous style, a calculating state-of-mind, a machinating *esprit.* The mercenary operator was the self-marketer practicing power plays, not persuasion. Like the wily Ulysses, he was *polytropo.*[100] Diabolically versatile, this "man of many turns" embodied sophistication or the abuse of rhetoric that selfish men employed to ingratiate themselves with gullible consumers and thus make their fortune.

Fénélon, tutor to the Duke of Burgundy, systematically attacked "fruitless eloquence that satisfies the vain curiosity of hearers and encourages their idleness." Such evil "Haranguers" did to the intellect what depraved cooks did to the stomach. Their rhetoric was only an art of dressing up delicacies

Kaspar Schott

*Demonstration of How to Make Water*

*Ascend One Side of a Hill*

*and Descend the Other*

1664

from *Joco-Seriorum Naturae et Artis*

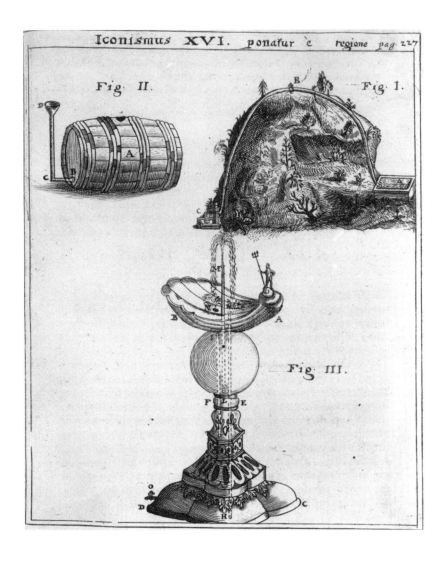

to gratify the corrupted taste of "Women and the undiscerning Multi-tude." Although orthodox in practice, Fénélon's quietist leanings were evident in his identification of such "feminized" insinuations with the charlatan's or quack's bag of tricks. The verbal conjuror "patches together Shreds of Learning" and relies on "little Knacks" to please and amuse. Rollin agreed in decrying the eroticization of the rhetorical process. "Decked out, tricked up and painted" discourse dazzled and deluded the hearer into acquiescence through surprise, "faint glitter," and "soft enchantment," appealing to "superficial minds" easily charmed by novelty.[101] Brilliant gusts of images remained synonymous for De Quincey with the apparatus of sophistry. This art of superfluous ornament, that is, of artificial or *technical* display, did not so much laboriously persuade as automatically delight. "Effeminate babblers" coerced involuntary assent from the idle through bewitching verbal extravagance.[102]

From the perspective of the Enlightenment, then, a deceiving "diamond dust" bestrewed the major institutions of the *ancien régime.* Remember that throughout the eighteenth century "pompous raiment," adulterated speech, and idolatrous imagery were identified with religious conjurations, usually Catholic, directed at the credulous.[103] Thus Hume railed against the "magnification" of grossest delusions when fools propagated "the imposture" of supernatural and miraculous phenomena.[104] Luxury, ostentation, indolence, and the rhetorical desire to show off were epitomized from Rousseau to Brewster in phantasmagoria. This pseudoempiricist demonstration spread "the singular illusions of sense."[105] Apparition, prodigy, and phantom had real social consequences when powerful technology was used to cheat through "shews and ceremonies."[106]

Decremps, in *La Science sansculotisée* (1792), criticized the abstruse and unilluminating style of certain "mystifying" scientists "speaking in an unknown language to [ordinary] people in order to dispense speaking reasonably with them." Resembling Freemasons, they wielded an obscurantist language and a confounding technology to exclude and baffle the profane. Demonstrating nothing, they simply demanded to be believed.[107] Similarly, Aimé-Henri Paullian, in his *Dictionnaire de physique* (1758), had warned against the rising sect of modern materialists. Comparing Hobbes, Bayle, Voltaire, and La Mettrie to *joueurs de gobelets,* this Jesuit professor of physics at Avignon maintained that atheistic philosophers made the same untenable claims and employed the identical dubious means as conjurors. Jugglers "through artful tricks and cunning take

94

Jean-Antoine Houdon

*Giuseppe Balsamo, Comte de Cagliostro*

1786

possession of the senses in order to beguile reason; they [the materialists] know how to distract and to present [to the intellect] as cause of an effect that which never was nor ever could be. Thus the latter, in their sophisms, speak only of the imagination, they address only it, and speak only in its name. . . . The imagination, that artful cause, they offer to distracted wits as the unique principle of all human operations." What clinched the analogy between the mountebank and the materialist, for Paullian, was the fact that both were "seducers without being seduced, diverted themselves through the credulousness of their stupid admirers."[108]

The self-asserting charlatan, like the grotesque-making fantasy, thus threatened mental, social, and linguistic constancy. Man and faculty alike possessed the destabilizing ability to collate disparate phenomena through optical projections. No case better encapsulated this multifaceted Enlightenment critique of artificial persons and fictitious combinations than that of Cagliostro (fig. 94). The crypto-raptures and automatic visions of this self-proclaimed Grand Cophta of the Egyptian Rite were compared to the virtuoso's trapping web, spun of lies and fantasies.[109] Like the Machiavellian man of *virtù*, this mesmeric intriguer knew how to seize opportunities and use them to his advantage.[110] Religion, politics, art, economics, science, and show business converged in the carnivalesque case of "this prosperous charlatan." Incarnating Derridean "différance," or the systematic play of differences,[111] Cagliostro epitomized impersonation. The devious life of this "irregular" Mason, Illuminist magus, and occasional Swedenborgian demonstrated that virtuosity was the opposite of virtue.

In Roche de Marne's pamphlet the *Mémoires authentiques* (1785), we are told that before coming to France he cheated merchants, tricked at cards, sold titles he did not possess in Russia, Prussia, Holland, and Austria. Subsequently ensconcing himself in the radical Whig circle around the Prince of Wales during the summer of 1786, this necromancer fled to Switzerland from London in 1788–1789. Cagliostro ignominiously ended his days in Rome, where he was chained to the wall as a revolutionary agitator by the Inquisition in 1790. Incarnating a truly international system of imposture, the scam artist embodied anticommunication by practicing an inverted pedagogy of hypocrisy. Mirrors, smoky rooms, and subterranean caverns formed the deluding apparatus and stagy sites for this *evocateur's* mysterious ceremonies. To his hypnotized adepts, Cagliostro admitted that the great secret of his art consisted in controlling the vulgar by never divulging the truth. "The tomb of St. Médard has replaced the shadow of St. Peter, the wand of Mesmer, the pool of the Nazarene philosopher."[112] Robertson, who argued for using phantasmagoria as a prod to enlightenment, remarked disdainfully that the outrageous life of this "great monger of prodigies" was nothing "but a series of *jongleries.*" Beyond conjuring ghosts in black-draped chambers through "the magic of optics,"[113] the great "sophister" was embroiled in one of the most dubious political events of the *ancien régime.*

The Affair of the Necklace (1785) was about disinformation, the deliberate creation of a false image. Moreover, this libel exposes the blurred edges of the eighteenth-century conception of play, capable either of edification or corruption. The stakes in this risky sport were high, since the purpose was to slur a "libertine" Queen whose supposed excesses were both erotic and economic. The often-told tale of Boemer, the court jeweler whose extravagant diamond necklace was rejected by Marie Antoinette and Louis XVI, was filled with a harlequin cast of calculating characters.[114] The unprincipled adventuress Jeanne de La Motte, an impoverished descendant of the House of Valois, duped Cardinal Louis de Rohan. This superstitious prelate had fallen out of favor and was seduced into thinking that the surreptitious gift of the necklace would reinstate him in grace. The "imposition" or "scheme" of La Motte, according to contemporary accounts, involved getting a young woman of the royal household to impersonate the Queen. This servant was to enact "a hoax," pretending to be Her Majesty frolicking in one of the garden *folies* located in a secluded corner of Versailles. There the false Queen accepted the desperate cleric's improper advances. For playing such a hazardous game, all parties concerned were eventually sent to the Bastille and ultimately tried before *parlement.* These included the Cardinal, Mlle. d'Oliva (the young woman who innocently acted the regal part), and Cagliostro.[115]

According to La Motte's vilifying *Mémoires,* this bestial "charlatan," "empiric," and "swindler" not only misled her but abused the power he had over the credulous priest. Significantly, the global "knavery" of Cagliostro was inseparable in the contemporary mind from the irrational "cabalistic seances" he confected. The anonymous author of *Cagliostro démasqué* (1786) noted that the conjuror had already been exposed as a fraud in 1780 while in Poland, well before beginning his Parisian "course of imposture." In a grotesque reversal of rational recreations such as Decremps's "proper amusements" for leisure hours,[116] this prestidigitator used young girls as mediums. Gazing into a bowl of water, they fell into a state of visionary trance. In a parody of popular science lectures, the operator trained children mindlessly to repeat mumbo jumbo while pretending to communicate with the Seven Celestial Angels. These tricks of optical illusion were enacted in a pseudolaboratory.[117]

Thomas Andrew James's historical and anti-Catholic novel *The Charlatan* (1838) functioned not unlike the scurrilous French pamphlets composed at the time of the Affair. It, too, presented a double satire of the hypocrisy and idolatry riddling the *ancien régime.* One of the key scenes was set at a *soirée* with Diderot, Grimm, and Condorcet in attendance. Beaumarchais, responding to a request from the assembled company, replied that it was impossible "to describe a rogue with so many disguises." He was "the Prince of Charlatans—the very Caesar of Jugglers," a "wonderful Alchymist" transmuting folly into gold, credulity into banknotes.[118] When one of the characters, the Count d'Ostales, finally visited Cagliostro in the rue Saint-Honoré, he was obviously disappointed. Clearly expecting to be ushered into a grotesque *Wunderkammer,* the skeptic beheld, on the contrary, no such showy excesses. He discerned "no stuffed crocodiles, or lizards, or flying fish; no black circles ornamented with the signs of the zodiac, nor even a parchment scroll inscribed with magic characters. There were none, in short, of the ordinary paraphernalia of a modern conjuror." Instead, as if in a perverse mimicry of the elegant economy and purposeful gravity ruling Lavoisier's austere laboratory (fig. 95), two simple tables supported chemical apparatus, electrical machines, and other philosophical instruments.[119]

In a biting attack on modern experimental science, James has the pseudoempiricist mock those altruistic inductive philosophers who, like d'Alembert, Lavoisier, Leibniz, and Newton, consumed their lives "in poring over isolated facts, and then timidly venture, towards the termination of their career, to publish a few partial, uncertain inferences." Scorning private study and self-effacing labor, Cagliostro's exhibitionist "course of lectures" ascended to abstract truth. If the objective of the "vulgar herd of [scientific] geniuses" was "to enlighten their fellow creatures," the technocratic virtuoso's goal was "to elevate and glorify my inner self, while I, in my esoteric practice, endeavour to mystify and degrade mankind."[120]

Resembling the episode in *Emile* when the child, during his first outing, encountered the insincere magician and his automaton, Cagliostro's phan-

95

Jacques-Louis David

*Antoine Laurent Lavoisier and His Wife*

1788

tasmagoria symbolized the shattering impact of fraud on the naive and vulnerable psyche. Rousseau's mourning over the "contemptible morals of the present age"[121] seemed to suggest that society's role in the manufacture of inauthenticity was actually minimal. Transgression was not just the sign of the times; rather, there was something innately crooked in human beings.[122] Despotic egotism, the gullible *amour-propre* of individual agents—so easy to manipulate, reify, and make public—disfigured civilization. Rousseau's combat against human nature's capricious unfolding called for a therapeutic pedagogy, one aimed against the sophistic production of "double men." Like the theatrical charlatan, false educators were pathological liars. Con men and nostrum-selling quacks were "always appearing to relate everything to others and never related anything except to themselves alone."[123]

This chapter has examined the system of imposture against which the enlighteners relentlessly battled. Indeed, I suggest that what permits us to speak of *the* Enlightenment is precisely the ubiquity and diversity of scheming in opposition to which the age's chief thinkers defined themselves. Shiftiness, wheeling and dealing, visual forgery, and verbal intrigue, as Rousseau intimated, were not unique to the *ancien régime*.[124] Yet early modern tricksters—taking advantage of the consumer's automatic acquisitive drives—contributed something special to the long tradition of gyp artists commemorated in the satires of Rabelais, Molière, Joyce, Beckett, and Pinter. We have seen how the concatenation of misleading appearances, deceptive optical technology, and the manipulation of knowledge served to shape the larger eighteenth-century critique of a "papist" oral-visual culture.

The specifically educational slant of the war against simulation was intimately tied to the logic and fantasy of games. Ambivalent and farcical vaudeville, that is, "juggling" in its broadest connotation, operated according to rules that mechanized the intercommunicative universe.[125] Cagliostro's dissimulations, as in a farce, turned his victims into helpless and ignorant puppets subject only to will of their manipulator. Similarly, the nonsense and make-believe typical of child's play obeyed a logic as controlling as that governing the magic act.

On one hand, the *invisible* or concealed imposition of artificial patterns could change the world into a toy and the players into tools of the hidden juggler. As in Cheselden's plates, objects could thus appear to exist "objectively," as if emitted "naturally" or automatically. On the other hand, the illusionist could make instrumentality *visible* by informing the viewer of the tactics used in the performance. Thus all exhibitionism, whether functioning within art or science, possesses this double inclination toward irrationality and rationality. Demonstration contains the potential either to shape action by concealed stratagems or to persuade by the revelation of the creator's protean means.

We shall see how this fundamentally aesthetic duality affected the staging of experiments during the Enlightenment. Like the acrobat and mountebank, demonstrators continually manipulated their bodies, used them as raw material. Self-transformation and the alteration of materials were thus part of the same system combining art, technology, and game. Making a malleable plaything of their own substance, experimentalists could not escape the onus of sleight-of-hand.

96

Joseph-Aignan Sigaud de La Fond

*Apparatus Demonstrating the Impenetrability of Bodies*

1775

from *Description et usage d'un cabinet de physique*

### The Art of Experimentation

The moral outcry raised against the charlatan highlights the crisis of handwork in early modern Europe. Juggling was a manual trade, much like experimental science (fig. 96), weaving, furniture making, or silversmithing (fig. 97). It, too, depended on a mechanical dexterity difficult to put into words. While it might appear that skill or artistry was unfeignable, we observed in the preceding chapter the frequency with which its processes and products were faked during the eighteenth century. The enlighteners believed with Plato that a profession, unlike a merely mercantile occupation, was an ethical act inseparably linked to the virtue of the performer.[1] The difficulty, however, was that any *technē* created some object detachable from the behavior, biography, and body of

97

Nicolas de Largillière

*Portrait of Mr. and Mme. Thomas Germain*

1736

the maker. Virtuoso know-how manufactured separate and superfluous goods. Alluring commodities, circulating uncontrollably in society through the agency of money, thus signified the antithesis of noble altruism. Gaining a livelihood meant treading the same path, reproducing for a fee and hence vulgarizing those items and experiences people needed or craved.

Serving the interest of others could, as in the case of the mountebank, also lead to their manipulation in the operator's self-interest. This corruption of manual practice because of a hunger for power or love of profit came about through disembodiment. Knack was a repeatable routine uprooted from its source in the character of the actor, and from the expectations and capacities of beholders who became mere objects for disguised, if expert, handling. This ambiguity haunting purchasable labor and, more generally, physical activity is still endemic to our late twentieth-century "service economy." The dichotomy between highly esteemed professionals, engaging in original brain work, and brutish toilers, plodding along in humdrum chores, was one of the legacies of the Enlightenment.

The "demon" in eighteenth-century economics was the tendency to see manifest execution as no more than a pretext for fiendish tricks. What Ernest Gellner called the linguistic "devil" of postmodern and poststructuralist systems actually derived from the early modern reification of bodiless reason. Schematic logic, judged capable of determining truth or falsehood, was erected as a bastion impervious to the blandishments of sensory misrepresentation.[2] The regulatory role of writing, method, and mind thus came to govern scientific and aesthetic practices since the seventeenth century, especially those prone to less predictable optical, gestural, and emotional forms of communication.[3] Any fine art that purported to demonstrate something—if only how to hold a puppy (fig. 98)—and any scientific experiment sustained by machinery—no matter how obvious the technique (fig. 99)—was liable to the accusation of *trompe-l'oeil* and pandering to spectatorial desire.

Significantly, the growing caste of professionals responsible for the later eighteenth-century theoretization of the crafts distinguished themselves from pedestrian "mechanics." Invisible quality of mind, not visible agility of hand, was requisite for the generation of impalpable ideas and an etherealized *je ne sais quoi*.[4] The ideological split dividing the eighteenth

98

François Hubert Drouais

*A Boy with a Black Spaniel*

1766

## 99

Jean-Antoine Nollet

*Experiment with Oscillating Pendulums*

*of Different Lengths*

1743

from *Leçons de physique expérimentale*

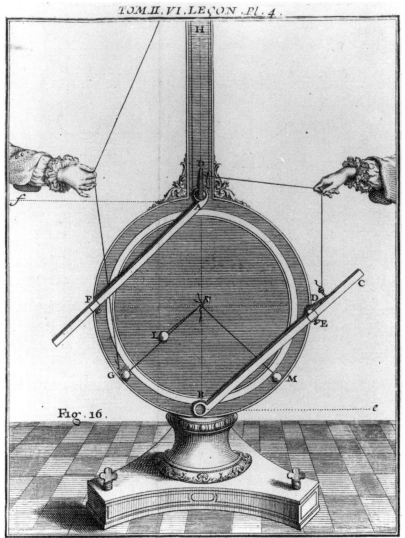

100

Nicolas de Largillière

*Portrait of the Sculptor Nicolas Coustou (1658–1733) in His Studio*

1710–1712

century is usually formulated in aesthetic terms as a "serious" neoclassicism vanquishing a "frivolous" rococo. This rift, however, might better be considered as the result of a much larger battle between material, devicelike sets of activities and mental or cognitive functions. On this point, Nicolas de Largillière's three-quarter-length portrait of Nicolas Coustou, flamboyantly displayed in his *atelier,* is especially telling (fig. 100). The rococo sculptor's gesture symbolizes sleight-of-hand: with a flourish, he changes a rude block into a tangible masterpiece without divulging his recipe for the transmutation.[5] Equally telling is Anton Raphael Mengs's unsparing rendering of his sickness-ravaged features (fig. 101). Shown conspicuously without the tools of his trade, the German painter, as in so many neoclassical artists' portraits, focused attention on the psychologically complex expression of the face. The head thus metonymically stood for the thoughtful inner life of the grave intellectual in contrast to the painfully acquired external maneuvers of the alchemist-artisan.[6]

In this chapter, then, I explore the prevailing attitudes toward materials and hands-on technology before the Industrial Revolution. The etiquette of observation and experiment, the body tricks of public science performers, the spectacle of automated toys, and the pantomime of businessmen's gestures belonged to a creative world of entrepreneurial promotion lying beyond the *Encyclopédie* and outside the walls of learned academies both in England and France. For inventors, whether in the visual arts or in natural philosophy, the exhibition of thought occurred in terms of concrete demonstrations and "mechanical representations."[7] Concepts manifested themselves in new arrangements, striking incorporations, and in cunning levers, screws, wedges, and pulleys. Experiments, inventions, and instruments allow us, therefore, to witness visual education in the making. Moreover, the artificial "event-object"[8] fabricated in the laboratory, on the stage, or in the studio revealed how complicated the process was for the moderns of bringing together the methodical with the material, theory with practice.

In *Body Criticism,* I argued that works of art were like biological organisms. This fundamental analogy was inscribed in a complex metaphorology. As physical entities, such products of manual labor could not be divorced either from their culturally conditioned conceptualizations or their material processes of embodiment. Similarly, the visible and tangible phenomena generated in the public space of the laboratory came into

101

Anton Raphael Mengs

*Self-Portrait*

1778–1779

being surrounded by a cluster of epistemological problems. Experimentalists were aware that messy observations had to be transformed into commonly accepted matters of fact. Data aggregates had to be hammered into unitary abstractions for a scientific community increasingly interested in quantification. Only by following the right rules of conversion could material images become symbolic representations and thus the objective foundations of scientific knowledge.[9]

Sebastien LeClerc's paired illustrations from the *Pratique de la géométrie* (1691) marvelously demonstrate how calculation might perfect an imperfect nature (fig. 102).[10] In analogical fashion, the subdivided figure floats above a singular rock mass, actualizing the boulder's inherent potential for triangular and pentagonal regularity. The broader message conveyed by the plates of LeClerc's small and portable manual was that a transcendent or underlying geometry made clear and distinct the troubling surface irregularity and entropy found in this world's landscape. More generally, his juxtapositional method persuaded the viewer of an axiomatic truth. While one times one equaled one, there was always a disjunction between the original and its copy (fig. 103). This abstraction was embodied in the gallant scene of women gazing into a basin of still water in the garden at Versailles. These courtiers discovered the human condition, symbolized by the mirror of Narcissus, namely, that the double was always only the far-off and blurred reflection of a remote and focused divine prototype.[11]

The ontological distance, allusively figured in LeClerc's engravings, separated the region of tactile things from the sphere of ungraspable ideas accessible only in a vision. This dualism corresponded to the two realities of eighteenth-century laboratory life. Borrowing the language of contemporary ethnomethodology, the dichotomy can be said to hinge on the tension between digitality and opticism.[12] Both the epistemological and the practical "problem of experimentation" was that it relied on the play of fingers (fig. 104) and the intervention of lensed instruments (fig. 105). The vexed question of how to transform the deft creator and perceiver of marvelous effects into an impartial *observer* of textualizable facts was thus of paramount importance to the scientific actors themselves.

There exists an untapped scientific literature wrestling with virtuoso dexterity, on one hand, and the master showman's compulsion for visualization and self-display on the other. The Dutch experimental physicist and instrument maker Pieter van Musschenbroek (1692–1761) initiated

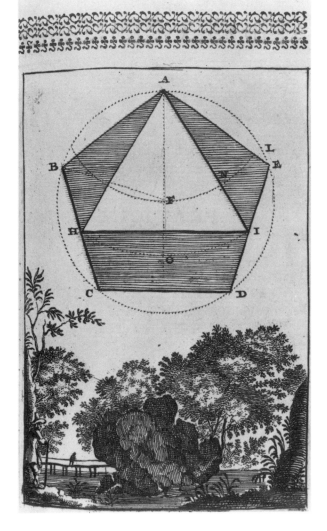

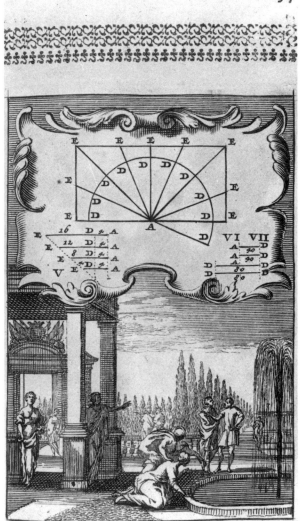

102

Sebastien LeClerc

_____

*Demonstration of How to Inscribe*

*an Equilateral Triangle into a Pentagon*

1691

from *Pratique de la géométrie*

103

Sebastien LeClerc

_____

*Demonstration That Things Doubled Are Equal*

1691

from *Pratique de la géométrie*

Adam Walker

*Magnetism*

1799

from *A System of Familiar Philosophy*

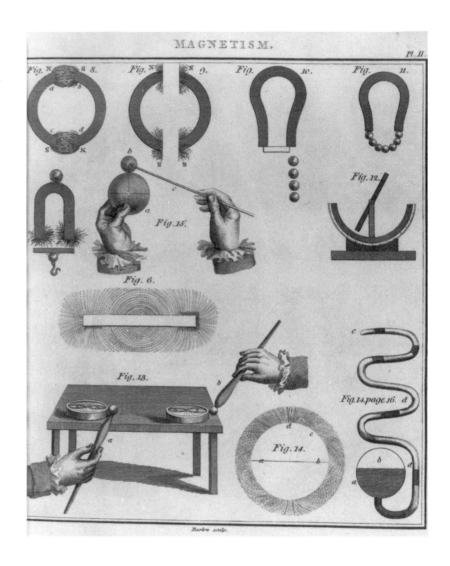

105

Nicolas-Bernard Lepicié

*The Astronomer*

1777

this series of important Enlightenment reflections on the complexities inherent to the demonstrative mode. The *Discours sur la meilleure manière de faire les expériences* (1736), frequently cited throughout the eighteenth century, struggled to define objectivity in the realm of manipulation and appearance. In addition to perspicacity, the experimentalist had to avoid exclusively embracing any particular system, whether that of Aristotle, Descartes, Stahl, or Newton. Partisanship, he argued, condemned the scientist to see things in a biased fashion.[13] Single-minded adherence to an ideology precluded the possibility of registering contradictory information. Further, Musschenbroek warned against classifying bonafide experimentalists such as Boyle, Torricelli, Guericke, Huyghens, and Mersenne—who checked their sensory impressions by submitting them to "right reason" and the corrective of multiple observations—among three types of pseudomechanics only apparently pursuing the same course. The fastidious observer, "délicat sur son travail," must not be confused with credulous opinion-mongers susceptible to "a thousand vague rumors." The act of seeing was not an operation. Unlike the anamorphic games of the Jesuits, scientists were not allowed ostentatiously to deform and manipulate images and to designate as facts obscure or ambiguous occurrences. Nor might they behave like cheating conjurors, bent on giving themselves "je ne sais quoi quel air de merveilleux," by theorizing from experiments they never carried out.[14] Like Salverte's necromancers toying with phosphorus, galvanism, and electricity, such modern magicians based their performances on spellbinding incantations, secret recipes, and mystifying prestidigitation impossible to replicate.[15]

Attentive to the "systems of imposture" riddling the old regime, Musschenbroek was keenly conscious of natural philosophy's potential for being subverted and appropriated by corrupt practitioners. He urged "subtle analysis, a scrupulous precision" to militate against capricious sightings, artifacts of instrumentation, and implausible findings published out of vanity. Microscopy especially, he felt, was prone to fantastic speculation. Even Louis Joblot, respected professor of mathematics in Paris, was not above reporting the sensational discovery of tiny animals with exotic human features, supposedly seen swimming in an infusion of royal anemones.[16] More generally, and resembling present-day witnesses to cold fusion, these "philosophes prévenus d'un système" or "prétendus méchaniciens" claimed they beheld something whereas they only imagined it.[17] The cult of public science, flourishing from the 1730s onward because of spellbinders such as the Abbé Nollet and William Whiston, encouraged virtuosi to ever more fanciful projections (fig. 106).[18]

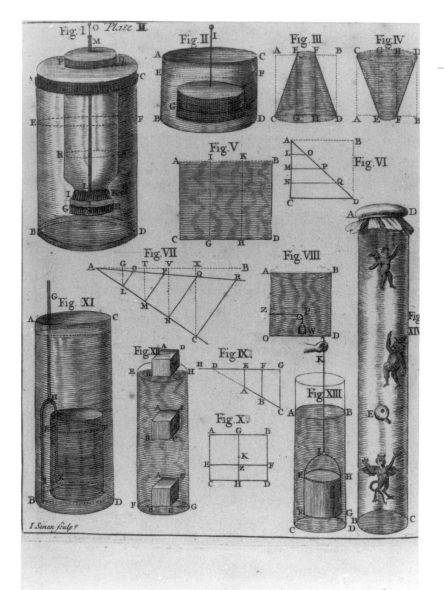

106

William Whiston

*Hydrostaticks*

1713

from *Course of Mechanical Experiments*

107

Bernard Forest de Bélidor

*Launching of Bombs*

1734

from *Le Bombardier françois*

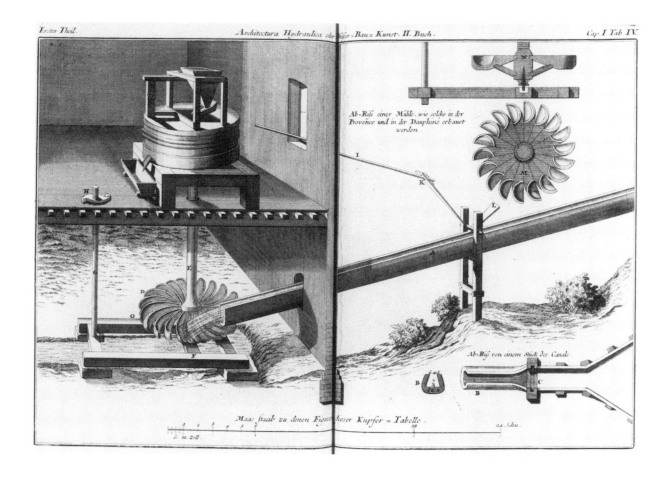

108

Bernard Forest de Bélidor

*Watermill Found in Provence*

*and the Dauphiné*

1741

from *Architecture hydraulique*

Like Musschenbroek, Bernard Forest de Bélidor (1693–1761) blamed the factionalization between "theorists" and "practitioners" for the misapplication and misunderstanding of technology. In his influential engineering treatises *Science des ingénieurs* (1729), *Le Bombardier françois* (1731), and *L'Architecture hydraulique* (1737–1739), the internationally renowned expert in ballistics and fortifications argued that mathematicians must hone their applied skills and that engineers, whether in peace or war, needed to study the rational principles governing their manual art (fig. 107). As Professor of Artillery at La Fère Military Academy, Bélidor had his feet in both camps. He believed that pure science also had to function within daily life. Conversely, empirical devices relying on handicraft, specifically hydraulic systems, could operate efficiently only if they were knowledgeably designed and demonstrated (fig. 108). Incomprehension of the basic elements of mechanics resulted in vague blueprints leading incompetent artisans to build watermills in which the fundamental laws of motion, flow, and friction were defied (fig. 109). Bélidor disparaged this neglect

109

Jacques Besson

*Horizontal Watermill*

1582

from *Theatro de l'instrumenti & machine*

110

Bernard Forest de Bélidor

*Demonstration of Friction*

1740

from *Architectura Hydraulica*

of know-how by mathematicians who identified research with handless theory and this woeful ignorance of the laws of mechanics by workmen whose unguided hands tugged on the cogs and pulleys driving useful machines (fig. 110).[19]

The absence of an intelligent and skilled labor force also impelled Nollet to write *L'Art d'expériences* (1743). Intended as a supplement to his *Leçons de physique* for professionals, this three-volume work addressed an audience of amateurs and, by extension, physics instructors in secondary schools. His aim was not to reproduce a collection of every conceivable instrument, but to enable the viewer-reader to fabricate his own devices or order them manufactured according to accurate specifications. A popular science demonstrator in Paris and, beginning in 1753, Professor of Physics at the Jesuit Collège de Navarre, Nollet cogently outlined a philosophy of experimentation. The preface counseled against superfluous apparatus, while urging durable construction and restraint in the application of ornament for easy maintenance (see fig. 99). An inventor of note (he improved Musschenbroek's Leyden jar), Nollet predicated his machine aesthetics on

a spare style.[20] Laboratory implements should be economical in having more than one application. Equipment ought to be not only accurate but simple, thus saving storage space in the crowded *cabinet de physique* or classroom. What was true of instruments also obtained for the art of experimentation as a whole. The scientist was urged to be ethical, to educate the public concerning the methods employed to achieve visible effects. This antisophistic tactic of initiating beholders into concealed causes stood in direct opposition to the artificial aids furtively manipulated by the con man.[21]

Bruno Latour, commenting on the sociology of present-day laboratories, asserts they are not merely the place where scientists work but mediated paper and technology worlds.[22] This perception of instruments propped up by publications is pertinent for the eighteenth century as well (fig. 111). The skeptical atmosphere of the laboratory was constituted of real or hypothetical observers, as in Cherubim d'Orléans's austere and subjectless camera obscura. Actual and virtual investigators watched as images were incorporeally extracted from monumental setups (fig. 112), then purged of "noise" by being focused and redrawn for display (fig. 113).[23] When analyzing the explosion of mathematical recreations in chapter 1, we remarked how the staging of playful and instructive games

112

François-Saneré Cherubim d'Orléans

*How to Observe Celestial Phenomena*

*(Sun Spots) in a Camera Obscura*

1671

from *La Dioptrique oculaire*

François-Saneré Cherubim d'Orléans

*A Newly Invented Instrument for*

*Drawing All Sorts of Objects in Proportion*

*(Front View)*

1671

from *La Dioptrique oculaire*

before a live audience left such performances open to the charge of juggling, examined in chapter 2. Like the chatty illusionist, the eloquent researcher purports to speak for equipment that does not speak. The silence of Cherubim's automaton eyepiece or the deaf and dumb barometers and plumb bobs, captured in Johann Wilhelm Meil's etching, seemed to demand ventriloquy (see fig. 111). Their intransigent muteness elicited the intervention of an acrobatic speaker to interpret inscriptions hieroglyphically registered on the hardware.[24]

This tension between demonstration as audio-visual spectacle and as the textual or numerical rehearsal of claims made about new objects escalated during the Enlightenment. The value of "a self-evident proof" was not the same as the "reiterated proof of some phenomenon."[25] More rigorous than either Musschenbroek's or Nollet's conversational books concerning proper laboratory practice, Benjamin Carrard's dialectical essay on *l'art d'observer* (1769) mapped out a martial strategy. He called for subjective vision to be disciplined by logic and corroborated by the testimony of the other senses. Discoveries were often no more than inchoate "collections of uncertain and poorly perceived events." His legal training was palpable in the emphasis on reductive analysis, the decomposition of sensory evidence, the injunction to examine data close up and to verify results through their exacting repetition. Carrard hoped that this "art of observation" might bypass the frailties and delusions common to individual sight by having many investigators systematically search for the same truth.[26]

Violent metaphors, reminiscent of the anatomy theater rather than the inorganic laboratory, incongruously colored Jean Senebier's *L'Art d'observer* (1774). This Swiss botanist and ordained minister dismembered the activity of willed seeing into three distinct temporal units: before, during, and after scrutiny. He stressed more emphatically than his precursors that observation was radically different from experimentation. The former entailed "a deliberate gaze" (*ce regard réflechi*) cast by natural philosophers on phenomena, like the double arc of the rainbow, to calculate their visible properties (fig. 114). The latter, as demonstrated in Newton's trials with the prism, was a probing method "for penetrating what escapes the senses, by forcing objects through artificial intervention to expose what they conceal" (fig. 115).[27] Although certainly more perspicuous than the unscientific but sensitive beholder alert to the changing look of ephemera (fig. 116), the rational observer only sees things as they are. Even when assisted by optical apparatus, he can only witness nature's ready-mades with greater acuity. The experimentalist, on the other hand, achieves cognitive clarity by placing materials in contexts not found in nature. Instruments contributed to this interrogatory process. Senebier cited the artifice of the dissector, wielding a scalpel to strip away skin and muscle (fig. 117), the mineralogist, grasping tools to dig into dirt, and the chemist, manipulating beakers, distillation vessels, and bell jars to decompose and recompose substances (see fig. 96).[28]

114

Johann Wilhelm Meil

*The Rainbow*

1765

from *Spectakulum Naturae & Artium*

115

Giovanni Battista Pittoni with Domenico and Giuseppe Valeriani

*An Allegorical Monument to Sir Isaac Newton*

1727–1730

116

Jean-Baptiste-Siméon Chardin

*Boy Blowing Bubbles*

c. 1735

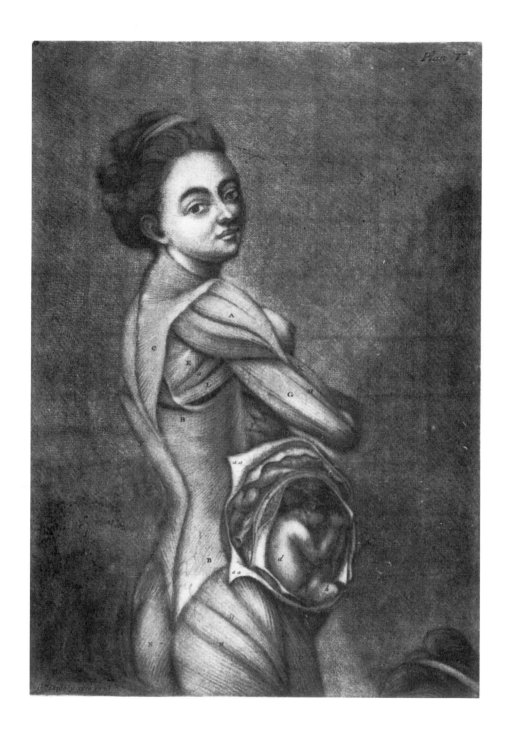

117

Jacques Gautier Dagoty

*Pregnant Woman at End of Her Ninth Month*

1773

from *Anatomie des parties de la génération*

In short, the observer simply "read" nature "like a book." In the manner of a lover ardently contemplating the beloved, he was loathe to alter a single feature for fear of mutilation. But the experimentalist did not yield to tender emotions. He tortured matter to expose its invisible secrets. Continuing the gender-determined metaphor, Senebier described the extraordinary measures required to compel physical phenomena to abandon their customary ways. The raw power of technology produced artificial combinations recalling not only Leibniz's ideal *ars characteristica* and Wolff's abstract *mathesis universalis* but the virtuoso legerdemain of fairground conjurors.[29] Senebier's "artful science," procured through "the games of our laboratories," contrived "a thousand extraordinary spectacles" and thus attracted cheats not adverse to falsification. He signaled the perils of rigged effects, "bizarre systems," and "ingenious novels" stuffed with doubtful data, to insist that observation must always precede application.[30] Perhaps he remembered the manipulative urge typical of seventeenth-century Baconians, all too eager to "twist the lion's tail," that is, to make clever objects of knowledge that were never found in nature.[31]

Importantly, Senebier developed parallels between proper scientific invention and the codes of decorum governing the fine arts.[32] He felt that experimentalists, whether working in the studio or the laboratory, should look to excellent masters for their models. Scientists and artists alike relied on visual persuasion, even while recognizing that varying levels of ocular proof or demonstration were possible. Not all observations attained the subtlety of Newton's, just as not all sculptors achieved the perfection of execution exhibited in the *Apollo Belvedere.* The nameless mason cutting stones for Claude Perrault's east facade of the Louvre, the potter Bernard Palissy who, although not a Horace-Bénédict de Saussure, taught naturalists about the history of fossilized shells unearthed in the Touraine, contributed to scientific progress while making only modest finds.[33]

Senebier's reflections on true versus false observation and experiment owed to the example of Johann Georg Zimmermann (1728–1795). This famous Swiss physician, who attended Frederick the Great in Berlin during his final illness, illuminated the close connection scientific rituals had in the eighteenth-century mind with the undisciplined routines of ignorant quacks.[34] Like Carrard's and Senebier's treatises, *Von der Erfahrung in der Arznen Kunst* (1763) is invaluable for understanding the socially constructed opposition between acceptable and unacceptable manual practices. It, too, revealed the extent to which the empirical sciences, applied mechanics, "clinical" medicine, and the visual arts were seen as cognate handiwork and judged according to the same aesthetic and ethical protocols. The "artist-observer" resembled the experimentalist in collecting data from nature and then perfecting it (see figs. 82, 83, 84). This artificial "improvement," however, separated the fine from the mechanical arts only by degree. Carpentry and cobbling also required dedicated research and informed skill. But, according to Senebier, the one-sided concentration on the theory of poetry and painting obscured this point. Thus it was common to think that originality was the province of the perspicuous interpreter, critic, or user rather than the maker. Like the student's or journeyman's copy, shoes were merely a tradesmanly record of the preexisting matter-of-fact of feet.

The masterpiece generated by a genius, on the other hand, defined itself in opposition to the menial reproduction.[35] Imitation, even when a *tour de force*—as in the clouds of steam evaporating above a silver tureen in the opulent still life painted by Anne Vallayer-Coster—supposedly made lesser claims to knowledge than imaginative brilliance (fig. 118). New entities and compositions were created under artificially induced conditions determined by the artful scientist or experimental artist. Gainsborough's virtuoso demonstration of knot tying is just such a conspicuous display of pictures as the unnatural product of mental inventiveness and manual skill. Set within a "wild" landscape suggestive of the haunts of Artemis, an unmarried young girl's attire is adjusted to expose the creator's ingenuity. By maneuvering the furry cloak so that it was about to slip off a virginal shoulder, the British painter ostentatiously created a situation whereby he might prove how it could be chastely caught just in time (fig. 119).[36]

The eighteenth-century anxiety over finding or making, copying or creating, displaying or demonstrating, provides an additional insight into the two kinds of antithetical communication we have been exploring. The opticism of the keen "artiste observateur" resembled the natural philosopher's zealous notation of countless details eluding ordinary apprehension (fig. 120). This ostensive mode, exhibited in Jean-Marc Nattier's *Portrait of a Nobleman,* was predicated on collecting and recording optical singularities that preexisted their gathering into a composition. Demonstration, in its digitality, conversely proved that experiments as well as pictures were deliberate products of labor forcing new phenomena to surface.

118

Anne Vallayer-Coster

*Still Life with Lobster*

1781

119

Thomas Gainsborough

*Miss Haverfield*

c. 1781

120

Jean-Marc Nattier

*Portrait of a Nobleman*

1732

A. M. Angelica Kauffman inc: e del: a Bologna 1765.

The Enlightenment ideal of progress was pictorialized as tireless doing. Not content to be passively showy, Chardin, Kauffmann, Gainsborough, Reynolds, and Wright of Derby (see fig. 76) *instructively* exposed how diverse things functioned by bringing actions and objects into unexpected and overtly contrived relationship. The meticulous braiding of hair (fig. 121) or the careful extraction of molten candle wax to seal a letter (fig. 122) rendered visible a physical process that would otherwise remain invisible. Like Nicolas Bion's personification of applied mathematics, drawing refined geometrical figures with a compass (fig. 123), Chardin's actors expertly handled their humble implements. A woman and her

122

Jean-Baptiste-Siméon Chardin

*Sealing the Letter*

1733

Usages des INSTRUMENS
de MATHEMATIQUE.

123

Nicolas Bion

*The Uses of Mathematical Instruments*

1723

from *Traité des instruments mathématiques*

124

Sir Joshua Reynolds

*Honorable Henry Fane with His Guardians, Inigo Jones and Charles Blair*

1766

servant—artificially removed from their natural habitat, the bustle of a Parisian salon—were set before our eyes in a moment of deft private performance.

Reynolds, too, subtly subverted the nonstrenuous conventions of politeness and civility whereby gentlemen (fig. 124) and ladies (fig. 125) simply were, but never *did* anything. Playing with contradictions, he passed easily from one register to another in the same painting.[37] The British artist conceived Henry Fane as the embodiment of the aristocratic ideal: elegantly idle and effortlessly nonchalant. Nonetheless, if action was ruthlessly banned from the center of this group portrait in favor of pose, it emerged incongruously in the attentive guardians shunted to the periphery. Toying with the stem of a wine glass, nervously folding a brocade waistcoat, and fondling the belly of a glinting flask with comic intentness, Inigo Jones and Charles Blair injected the complexity of club and coffeehouse interaction into the complacency of an out-of-door conversation piece.

Nor were collective rites the province of men alone. Compared to Jean Raoux's play-acting and *précieuse* vestal virgins—barely touching the sticks meant to stoke the sacred fire (fig. 126)—Lady Bunbury firmly gripped a Herculanean cup. This mock-heroic gesture wittily embodied her determination to be archaeologically correct while sacrificing to the graces (see fig. 125).[38] In sum: there was a growing sense in the second half of the century that art, like natural philosophy, was visual education. As a form of cultural expression, painting was not only capable of making progress but could offer models for progressiveness. As a type of optical demonstration, it, too, was the product of trials, manipulations, and empirical processes.[39]

This performative notion that all human artifacts were the result of physical effort was at odds, however, with a concurrent immaterial minimalism. Recall that, for Hemsterhuis, the beautiful was that which "gives us the greatest number of possible ideas in the shortest possible time."[40] Etiquette also determined this abstract language of spatial and temporal simultaneity (see figs. 102, 103), diminishing contours, and quintessenced concepts. The aggressiveness of the neoclassical disciplining of matter and materials was grounded in the fear that overt handling veered dangerously toward cunning maneuvers and macabre self-display. The technologist, compulsively manipulating machines, violated the rational

125

Sir Joshua Reynolds

*Lady Sarah Bunbury Sacrificing to the Graces*

1765

126

Jean Raoux

*Vestal Virgins*

1730

norms of self-restraint and decorous effacement shaping codes of civility and *honnêté* (see figs. 96, 99, 104, 106, 107, 110).

Yet it is important to realize that the romantic preference for artifice and for an "optimum" ideal form alien to nature, as well as its infatuation with morbidity, derived from Enlightenment experimentalism.[41] Significant, too, for the nineteenth-century construction of the dandy[42] was the popular science demonstrator's contortion of his body to capture the attention of both high and low social orders. We need to uncover those melodramatic laboratory games, or high-risk athletics, characteristic of the darker side of the Enlightenment. A type of *pantomime parlée,* dangerous scientific gymnastics convulsed individual physiology.[43]

Recalling Madame Tussaud's grimacing wax heads of guillotined revolutionaries or Alibert's dermatological monstrosities housed in the Hôpital Saint-Louis, the electricians' face and limbs were distorted through electrical impulses.[44] The biological organism became mechanized as a sublimely victimized automaton or a Promethean creative instrument. Like E. T. A. Hoffmann's beautiful but uncannily immobile Olimpia from the *Sandman* (1816), the viewer is meant to be repelled or enthralled by an encounter with the technological supernatural.[45] Looking ahead, an antihuman thread leads from Enlightenment self-abusing scientists, to Dr. Coppelius's artificial daughter, to Hans Bellmer's excruciatingly disjointed submissive *Poupée* and Salvador Dalí's docile *êtres-objets,*[46] to the cyberpunk of contemporary science fiction whose self is in danger of being absorbed into network circuitry.[47] Anxiety and celebration continue to surround the making of flesh into machines and metamorphosing machines into flesh. Physical shock stands for the assaults daily inflicted on our nerves by the instrument-altered environment.[48] Jolting consciousness has become the essence of postmodern experience.

### Body Tricks

The body as teaching tool made an occasional appearance in mid-seventeenth-century mathematical recreations (fig. 127). But Kaspar Schott's humorous chain of gymnastic bumpkins, fast-talked into literally testing the depth of a well, were a far cry from the strange laboratory "schools" of the eighteenth century. Eccentricity, eeriness, daring, and a fascination with technology's ability to shape and control a world that was not a

Kaspar Schott

*How to Determine the Depth of a Well*

1664

from *Joco-Seriorum Naturae et Artis*

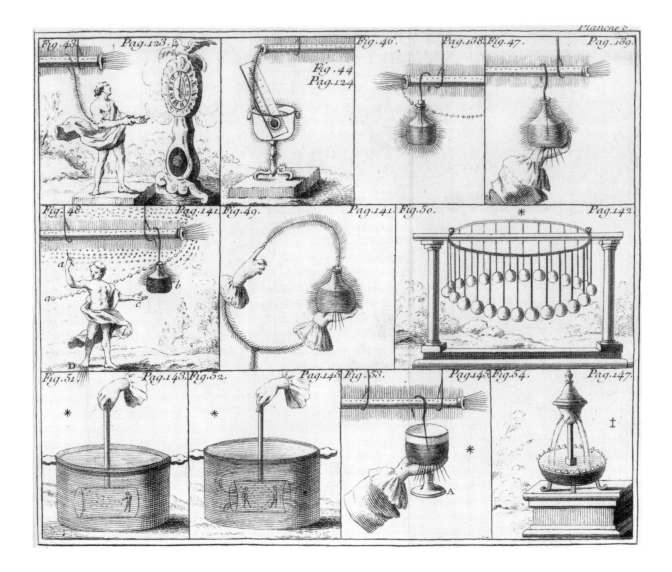

128

Charles Rabiqueau

*Electrical Self-Experimentation*

1753

from *Le Spectacle du feu élémentaire*

replication, but a bizarre parallel reality, reveals the grotesque side of enlightening entertainment (fig. 128). If the American poet John Ashbery is right that "recklessness is what makes experimental art beautiful," we can readily understand Nollet's cautionary admonition. "A course in experimental physics . . . [is] not a course in experiences."[49]

Dazzling operators, masochistically sacrificing themselves to weird procedures, risked losing sight of the instructive point in the heat of performance. Simon Schaffer recently dealt with certain theatrical rites in which the experimenter's body was presented as a form of self-evidence.[50] I want to extend his fruitful notion of staged science to consider its broader aesthetic, ethical, and pedagogical implications. Uncovering the ambivalent character of this vaudevillian troupe will also allow me to situate phantasmagoric electricians within the farcical, and sometimes horrific, Enlightenment project of popular education through sensationalist spectacles.[51] One of my major aims in this book has been to illuminate the fuzziness of the fine line separating conjuring ploys from the paroxysmic raptures of natural science in which the self, too, became art. Remember how Rollin warned his young readers against producers of showy signs. Such *tour-de-force* juggling was calculated to parade the genius of the rhetor, not to lay out information neutrally or disinterestedly. Hence exaggerated individualism was susceptible to the charge of mannerism and fraud.[52] We have seen in the preceding chapters how instructive games, because of their capacity to entrance the beholder, were interpreted both as the source of visual education and the excuse for manipulative display.[53] Stunning proofs of art or skill, whether emanating from the studio (see figs. 97, 119, 121, 122, 124, 125) or the laboratory (see figs. 96, 99, 104, 110) pleased through the captivation of others (see fig. 72).

The experimental environment, dependent upon elaborate setups requiring acrobatic dexterity, disturbingly resembled the trick hardware of the mountebank and similarly called into question the reliability of ocular testimony (fig. 129).[54] Moreover, the demonstrator's mesmerizing absorption in his own somatic functions proved difficult to distinguish from the flimflam merchant's mock pain and fake amputation. Michael Fried's convincing argument that Courbet's intuition of his own embodiedness was inextricably tied to conveying experiences that could not be put into words is pertinent, I believe, for the sensualism of eighteenth-century laboratory heroics.[55] Musschenbroek recounted how he used himself as a

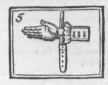

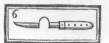

129

Henri Decremps

*How to Pierce Oneself in the Arms*

*and Stomach with a Knife*

1791

from *Codicile de Jérôme Sharp*

guinea pig, enduring the excruciating bites of venomous snakes and poisonous insects to investigate their harmful effects. Like Angelika Kauffmann's palpable fingering of her braid as if plaiting a flower wreath,[56] the Dutch inventor of the Leyden jar found it useful to examine his corporeal condition and test the stretch of his nerves. In advance of a trial, he always checked the look and feel of his hands, which tended to become transparent when suddenly heated.[57]

Belonging to the same artisanal proletariat, impersonators, purveyors of physical adventure, and makers of instruments were first and foremost competent workmen demonstrating special effects. As a group, they revealed the performative character of creative identity constructed in the sphere of public culture and outside the norms of controlling academies.[58] Nollet's or Rabiqueau's maverick "electricians" were showmen, indecorously exploiting the resiliency of their physique. Analogously, Nicolas-Bernard Lepicié (1698–1755) ostentatiously personified the matter of his materials, making them inseparable from his body. This rococo specialist in manufacturing lowly portraits and genre paintings capitalized on the tangible similarity between talcum, dropping from a wig and bespattering his waistcoat, and pastel. Ephemeral dust thus incarnated the craft of drawing, concretely captured as powdered pigments deposited on textured paper (fig. 130).

In the second half of the eighteenth century, because of pedagogical reforms French professors of natural philosophy were at last able to perform experiments themselves. These demonstrations, however, were not integral to the teaching of physics but tacked on, either at the end of a course or outside of classroom hours. The absence of illuminating visual aids, of sophisticated models and clarifying illustrations—beyond the diagram and blackboard and chalk—was lucratively addressed by a generation of popularizers feeding the consumer rage for "galant scholarship." England, because of the comparatively small number of its universities and Dissenting academies, also witnessed an explosion of itinerant lecturers to fill the visualization vacuum (see fig. 76).[59] Staged science, playing with mysterious forces and artfully deploying intriguing equipment, thus successfully marketed learning by packaging it to look like the thrills delighting audiences at old-regime magic shows (see figs. 47, 72).[60]

It was within this competitive entrepreneurial milieu, composed partly of hokum and partly of redemptive didacticism, that the independent natu-

130

Nicolas-Bernard Lepicié

*Self-Portrait*

1777

ral science demonstrator hawked both his person and his wares. Spellbinding performances challenged the dry textbooks and imageless dictations typical of the *collège de plein exercice* and the British secondary schools. As live conductors of gruesome fluids and disfiguring currents, bare torsos and exposed limbs made it seem as if sublime natural phenomena spoke directly to rapt beholders. Grimaces, twitches, and flailings defied discourse by displaying cosmic energy at the moment of its unleashing.[61] On one hand, then, the performer's exquisitely responsive physiology, like a delicately calibrated instrument, passively exhibited the objective power of another (fig. 131). On the other hand, the primal potency harnessed by the "new Prometheus" dramatized his own demonic vigor.[62] Feminine in his supple malleability and masculine in his aggressive boldness, the popular science demonstrator exuded and exerted universal appeal.

John Theophilus Desaguliers's (1683–1744) checkered career as a promoter of Newtonianism and as an equally avid venture capitalist was both typical and atypical of this peripatetic occupational group. Larry Stewart has suggested that his professional profile embodied the larger eighteenth-century transition from the intimate world of private patronage to self-employment and, finally, hire by public companies.[63] A Protestant born in La Rochelle, his family migrated to England after the revocation of the Edict of Nantes. Taught first by his father who ran a private school in Islington, he matriculated at Oxford after having shown an early bent for natural philosophy. Desaguliers's academic laurels (he succeeded to John Keil's chair in physics) were unusual among the later generation of electricians and mechanics, many of whom can be identified only by their last names and addresses recorded in newspaper advertisements. His importance lies in being an early disseminator of Newtonian concepts both as a demonstrator for the Royal Society and as a Drury Lane–type lecturer. George Rousseau, in a seminal article, described the formation of Desaguliers's private academy in Little Tower Street where, between 1712 and 1717, he delivered popular science talks to writers, politicians, Dissenting ministers, and medical men, and also sold his lectures as printed manuals.[64] Brilliantly successful in hawking information within the new natural philosophy societies or neighborhood book clubs—the "penny universities" that convened in coffeehouses—Desaguliers took his show on the road. Traveling to Holland, he hobnobbed with Ruysch, Boerhaave, and 's Gravesande.

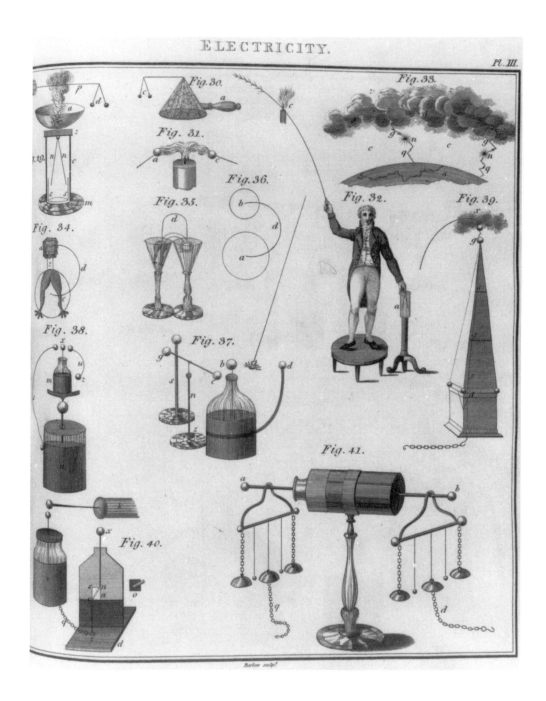

131

Adam Walker

*Electricity*

1799

from *A System of Familiar Philosophy*

Claiming there were only "eleven or twelve persons who perform Experimental Courses at this Time in England, and other Parts of the World," he boasted that he had trained eight of them by himself. Unlike "Quack Engineers," pushing apparatus of their own construction, Desaguliers relied on professional technologists such as Thomas Savery to supply equipment for various demonstrations. The design of steam engines became increasingly important in the mining ventures of the early eighteenth century, beset by the danger of noxious fumes and pit flooding.[65] After 1783, they began to be used extensively in the textile industry, replacing the waterwheels and horses that had made earlier factories no more than glorified workshops (see figs. 108, 109).[66] Consequently, the elucidation of Newcomen engines and other hydraulic contrivances was not just an amusing touch in a public lecture but a claim for special mechanical expertise and a call for technical training (fig. 132).

Beyond the obvious utility and visual appeal of clattering valves and hissing boilers, his popular talks were attractively accessible in asking only common sense from the spectator "and very little Arithmetic."[67] Desaguliers believed experiments could overcome the audience's lack of skill in mathematics. Although men and horses were required to carry out one physically arduous problem in applied hydraulics, no creature was shown wrestling with copper receivers or sucking and forcing pipes. Instead, bodiless automation reigned in John Mynde's engraving, which recorded the heroic presence of pumps, pistons, and the immediacy of the experiment.

Desaguliers, however, could also pointedly embody his explanations for an uncritical audience often conned by pseudodemonstrations. Drawing on a fascinating feature of popular culture, he subjected the engineered postures of "knavish or ignorant Pretenders" to scientific scrutiny. Exposing the mechanics and machinations behind street entertainment, he analyzed the poses of the "Kentish Fellow, Joyce" and "the German," who exhibited amazing acts of strength in the Hay Market. Relying on a psychophysiology common to electrical performers later in the century, Desaguliers transformed science into a physical sport. By alluding to a common sight in London's consumer-oriented society, he immediately engaged a broad audience that unabashedly enjoyed such spectacles but did not necessarily reason about them. Unmasking the "pretended Sampsons'" apparently automatic routines was thus a way of pleasurably capturing mass attention while surreptitiously slipping in an educational

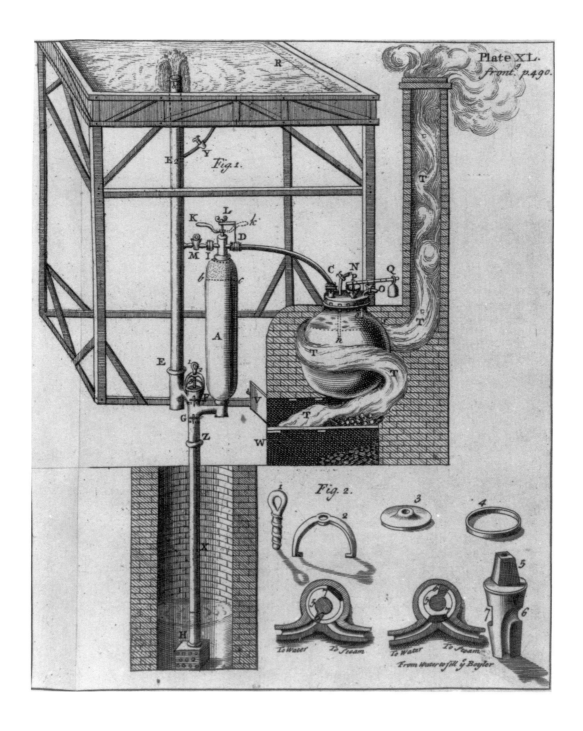

132

John Theophilus Desaguliers

*Steam Cylinder and Boiler*

1745

from *Course of Experimental Philosophy*

message. In the hands of a skillful instructor, these living teaching aids helped the beholder to experience abstractions empathetically in the form of muscles craftily lifting great loads.[68]

One century before Hazlitt, then, Desaguliers recognized the fundamental analogy between mechanical efforts and "actual experiments, in which you must either do the thing or not." Intellectuals can fool the public with their dogmas, but a "juggler or knife-thrower cannot persuade the audience at the Olympic Theater that he performs a number of astonishing feats without actually giving proofs of what he says."[69] But here was the rub. All forms of experimental display existed ambiguously between conjuring knacks and scientific know-how, between theater and learning. Self-inspection was crucial to both. The juggler, according to Samuel Rid, must cultivate a good grace and elegant carriage "to make the art more authenticall." Deportment, and even anatomy, helped to avoid bungling since body tricks "must be nimbly, cleanly, and swiftly done, and conveyed."[70] That "cunning handling" and "charming words" were also the staple of the philosophical illusionist was attested, as we saw, by the proliferation of treatises on the *proper* art of experimentation, stretching from Musschenbroek to Senebier. There is no smoke without fire! Think, for example, of the long sections found in that important eighteenth-century genre, the dictionary or encyclopedia of edifying amusements, devoted to how one might etch skin with acid or write invisibly on the scalp. Was this painful practice essentially different from Bion's tale of microscopists drawing their own blood to stare at under the lens?[71]

Both the magician and the experimentalist believed in incarnating thought. As De Quincey remarked, a great performance created a community through the demonstrator's artistry in synthesizing image, idea, and feeling.[72] Their *style* of incarnation equally veered toward the virility of the Longinian sublime: "that extraordinary, that marvelous which strikes in discourse and gives the work that force which ravishes and transports."[73] For the eighteenth century, the seemingly spontaneous shock of sublimity irrationally seized the unresisting body of the performer (see figs. 88, 94, 128). From this perspective, emotive enthusiasm, as the unbridled display of ingenious thoughts paraded "for the show of it, in order to satisfy the auditor's expectations," could also be connected to gendered rococo notions of a sensual and rapturous femininity.[74] Loosening the passions of others through ravishing sights was not, in the words of the Jansenist Rollin, "manly, noble, chaste" (fig. 133).[75] Indeed, the

133

Jean-Honoré Fragonard

*Blind Man's Buff*

c. 1750–1752

electrifier, bound to his flaming jars and exciting cylinders, resembled Fragonard's artfully disheveled and blindfolded young woman, provocatively tickled by a phallic fishing rod. Both were out of control of their person (see fig. 131). Mary Sheriff has taught us to see the scarcely disguised obscenities typical of the rococo pictorial code.[76] Extending this *aperçu,* I suggest that the French painter's nubile game player and Walker's or Rabiqueau's showmen have turned themselves into eroticized automata, into mechanized artifacts for the viewer's delectation. Yet, paradoxically, the experimenter's disciplined body was also akin to the regimented physique of the male slave to fashion (fig. 120) and to the aristocratic marionette of complacency (fig. 124). In the case of the popular science demonstrator, however, it was instruments, not apparel or attitude, that softened the flesh (see fig. 128).

In contrast to a theory-centric epistemology that provided an account of science as disembodied testing of mathematical propositions, staged experiments made discovery happen before the eyes. Desaguliers, Whiston, Nollet, Ferguson, Sigaud de La Fond, and Walker shared with rococo artists and artisans the practice of marking their material production with visible traces of the maker's self (see figs. 97, 100, 104, 116, 130). This corporealized knowledge was manufactured through maneuvers carried out within the organic environment of the laboratory or even among the meteorological sublimities of the external landscape (see figs. 96, 99, 106, 131). Beginning in the 1780s, a host of shadowy entrepreneurs rented *loges* in the wooden arcades of the Palais-Royal to trade in fashionable novelties.[77] Cheek by jowl with Dominique-François Séraphim's *boutique* projecting *ombres chinoises* was Sieur Pelletier's *cabinet de méchanique, physique, et hydraulique* with its ingenious machines.[78] Around and about town, the stroller might encounter the enlightened magician Pinetti, lionized by Decremps, conducting experiments in physics at the Variétés Amusantes. Nor was the famous circumnavigating botanist Michel Adanson above exhibiting his natural history cabinet near the Beaujolais.[79]

Charles Rabiqueau's popular science spectacles, presented in his *cabinet de physique* on the rue Dauphine near the Pont-Neuf, were typical of this later eighteenth-century charivari. Arguing for an embodied psychology *avant la lettre,*[80] he recounted in the *Spectacle du feu* (1753) how the publication of Benjamin Franklin's letters inspired the installation of a lightning rod on his house. On June 10, 1752, Rabiqueau dramatically unsheathed this fashionable instrument, running 90 feet from ground floor to rooftop. A copper filament extension was snaked into his laboratory "to satisfy the curious." Responding to the insatiable appetite of novelty-craving Parisians and to the manifest competitiveness of the electrical marketplace, he sequestered himself in his chambers to search for fascinating and familiar experiments "that might speak to the eyes in order to be understood by the whole world."[81] Attendance at this performer's well-equipped show—a bizarre *sons et lumières*—was likened to going to "a public library" by people who could not afford to buy books. Spectators were diverted by an electrified doll, the *Baccha Veritatis,* whose hair could be made to stand on end in fright. Her head nodded in approval, or struck the unfortunate supplicant foolish enough to make an impertinent request.

Rabiqueau, about whom virtually nothing is known except that he was one of the successful "amateurs en nouveautés & expériences soutenues," clearly borrowed the mountebank's carnival tricks. Like Rousseau's charlatan, he electrified a bowl of water on which he mysteriously attracted and repelled a toy boat. To great wonder and amusement, no doubt, he also charged the quills of a porcupine and stiffened limp ribbons into erectness by the touch of a wand. Playfulness soon became perilous, however, when the electrician turned high-voltage glass globes and energized wires on himself (see fig. 128). Putting his scantily clad body sublimely at risk, he stood on a wooden platform and gripped a chain leading to a conductor while electrifying himself in front of a clock. Unlike the childish games with metallic toys and iron fillings, the showman now deployed the cosmic fire to count the beats per minute of his racing pulse during shock therapy. When extending these physical trials to members of the audience, he discovered that men "without imagination" experienced no effect whereas others, even after the briefest contact, cried out in pain.[82]

Significantly, two decades before Mesmer's pan-European successes with the hypnotic *baguette,*[83] Rabiqueau had already described how both sexes, closeted within a darkened room, were "feverish to approach the electricity." This psychophysiological atmosphere of hysteria and hallucination was fed by the sulfurous smell of the "esprit de feu" penetrating the blood and rushing to the brain of impressionable participants. Turning the magical power on himself merely by pointing a finger or touching a bare leg, the demonstrator drew sparks from a phial. The resulting violent commotion convulsed his limbs before the flashes finally escaped through the feet. Success bred even greater recklessness. Like later galvanists and animal magnetists, he rubbed bottles to excite them and seized Leyden jars, inducing a burning discharge that reverberated in the surrounding air and contracted muscles into spasm. Such sensationalistic and even sadistic visual tactics, he felt, made the assembled company see and feel "the force of the aerial spirit."[84]

From a surviving handbill published June 17, 1772, we learn that this imaginative Parisian showman continued to thrive within the cutthroat Enlightenment entertainment business. Now operating at the sign of the "Grand Druide Automate," the entrepreneurial Rabiqueau had considerably expanded his changing multimedia illusions. Twenty years after his initial, lucrative foray into electrical experiments, customers were so plentiful that he could afford to relocate to the Hôtel Carignan on the rue Bailleul near the Louvre. This fashionable dwelling possessed a gateway large enough to accommodate a procession of "voitures Bourgeoises." Apparently they rolled by in droves to witness a sort of occult robotics, the "spectacle méchanique, *in supremis.*"

The advertisement offers a rare glimpse into the stylish world of old-regime science *à la mode.*[85] Seances occurred daily at 6 P.M. Holidays were no exception, not even religious "Fêtes solemnelles." The weekly cycle of amusements commenced with *La Fée Agenonié,* debuting Sundays and repeating on Thursdays. This dramatized fairy tale, set in the village of Dompré in Lorraine, took place in a tree where an ethereal band gathered to sing songs. The lead fairy underwent a series of enchantments by "Pyrotechnie hiéroglyphique," talismanic prestidigitation, and chemical powder of projection. These spells eventually landed her in a strongbox under lock and key. Monday's representations boasted cunning automata exceeding those of the legendary Vaucanson. Clients were assured of technological progress as last year's sensation, *La Perdrix rouge ingénieuse,* gave way before the *Limonadier ambulant.* The singing partridge had been supplanted by the perambulating lemonade seller, improbably accompanied by a winged Mercury serving as interpreter. Wednesday's performances were devoted primarily to magic acts: *Les Six salles exagones, Le Vaisseau, Le Tantale, pièce nouvelle,* and *Le Tableau magique,* also included in Sigaud de La Fond's rival repertory (see fig. 88). The midweek high point, however, was palingenesis. *La Fée* exhibited the resurrection of a bird from ashes to have it miraculously appear, fully fleshed, trapped in a cage. Resuscitations were evidently so popular that Rabiqueau repeated them on Sundays and Tuesdays. Saturdays were devoted to optical sports, especially catoptric tricks. Clients hungry for *nouveautés* could also engage in horoscopic games and in electrical demonstrations, specially devised for those ignorant of this fascinating phenomenon.

Popular science education did not come cheap. The first show cost three livres, the second, one livre, four sous. Twelve livres could register the auditor for the entire suite of thumping events. An indefatigable performer, Rabiqueau assured the public he would "represent" at any time, providing the party booked in advance for a slot falling outside the usual "heures des Spectacles." In slack periods, he was even willing to draw up astrological charts for twelve sous if five people made it worth his while.

These rotating diversions in no way exhausted the multiple pleasures of the *Cabinet privilégié du Roi.* The capacious Hôtel Carignan also housed a more high-minded *Ecole de Récréations physiques méchaniques & mathématiques* in which fairground charlatans'—that is, his competitors'—sleights-of-hand were unmasked. The demonstrations conducted in this "School" were made available in a "textbook." For 30 francs or three pounds sterling, subscribers to the *Livre de Récréations physiques & mathématiques* could find additional practical lessons in how to become wary consumers, wise to the craftiness of *Farceurs.*

Even more edifying was the *Cabinet*'s third objective: its weekly course on experimental physics. Everything elegant company needed to know about natural philosophy was crammed into six days. The syllabus is tantalizing. Mondays and Tuesdays focused on a variety of electrical displays. Rabiqueau pointedly remarked that neither the ingenious equipment nor the subtle operations could be witnessed elsewhere. Education continued apace on Wednesdays and Thursdays with lectures examining the properties of air. Handling pumps, fountains, compressors, and siphons was taught. Fridays and Saturdays were given over to mechanics, the manipulation of levers, pulleys, screws, and wedges. The instructor relentlessly and indefatigably moved on to a consideration of optics, dioptrics, catoptrics, and magnetism. These wondrous exhibitions, "as curious as they were satisfying," could be audited for a mere 24 livres. Nothing, it seems, was left to chance. Gentry might attend either morning or evening classes since the master had thoughtfully installed special *lampes optiques* to prevent eye fatigue!

Rabiqueau's violent self-electrocutions, then, were part of a gamut of corporeal and instrumental antics destined to amuse and illuminate wealthy patrons. The prospectus outlining the *Cabinet*'s special effects was a *chef d'oeuvre* of later eighteenth-century marketing and indicated just how popular these visual shenanigans were. Steven Spielberg's *Jurassic Park* is a modern-day reminder that scientific motives, themed entertainment, and making money have long been intertwined. Clearly, unabashed self-merchandising and self-promotion through frightening contortion sells. Seen against the background of such macabre science showmanship, Franz-Xaver Messerschmidt's playfully distorted self-portraits from the 1770s no longer seem odd or perplexing (fig. 134). This brilliantly executed series of 12 sculpted heads should be interpreted as a form of facial acrobatics. Like the corporeal *grotesqueries* of the medicine man, tumbler, or Pont-Neuf charlatan, they appear less the pathological work of a sick man,[86] and more part of the apotropaic phantasmagoria characteristic of the later eighteenth-century dusk of unreason. The Austrian artist, in the manner of the French and English electricians, studied the gymnastics of disfiguring involuntary responses to extreme physical stimuli. In a kind of self-autopsy, he, too, turned his body into a specimen. Like the bizarre grimacers, performing wordless farces at fairs, such ostentatious and *controlled* self-experimentation was blatantly indecorous and spoke to the people in a way that the *philosophes* could not. In advance of Gall's plaster skulls, localizing drives and capacities within specific regions of the brain, Messerschmidt's analytic busts exist somewhere in between entertainment and diagnostic tool.[87]

The old-regime fascination with magic, illusion, and science thus flowed unimpeded into an outlandish romanticism, equally smitten by electrical charges, moving gadgets, freaks, and funambulism. The rage for gory spasms witnessed in the morgue, on the gallows, and in the aftermath of the guillotine similarly confounded the supernatural with the somatic. Goya, Delacroix, Géricault, as well as Robertson, Pixérécourt, and the young Hugo relished *grand guignol* visions. Paintings and novels during the first two decades of the nineteenth century likewise served up a curious mixture of scenes culled from the laboratory, hospital, studio, and sanctuary.[88]

The materialism of Mesmer's electrical effluvium, bathing all creation, and the corporeality of Messerschmidt's psychic demons, supposedly exorcised through physiognomic routines, are apt on another score. The

134

Franz-Xaver Messerschmidt

*The Yawner* (model)

1778–1783

Catholic Rabiqueau also spoke the language of tangible mysticism when he described fiery fluids animating the marvelous "container" of the universe. Practicing a scientific incarnation theology, he viewed the micro- and macrocosm as an interconnected dynamic system. God was the ultimate experimentalist whose celestial fire animated all earthly bodies.[89] Joseph Priestley (1733–1804), too, believed in an indwelling principle in matter that needed to be brought forth. "The present race of electricians" imitated "in miniature all the known effects of that tremendous [heavenly] power . . . drawing the lightning from the clouds into a private room." The British Dissenting preacher, political writer, associate of the Birmingham Lunar Society, and chemist-rival of Lavoisier shared with Rabiqueau the craftsmanly tendency to picture phenomena involving light in quasi-substantial terms as "torrents," "particles," "vapors," and "vibrations."[90] He, too, was motivated to advance enlightenment by encouraging the public to participate in the enactment of natural knowledge conceived in terms comprehensible to them.

Adam Walker (1730?–1821) appears to have been a disciple of Priestley's and among the first to have included the chemistry of "airs" alongside astronomy and experimental physics in his repertory of popular lectures.[91] We learn from his own account that he was a highly successful philosophical lecturer who moved to London from Manchester around 1778, where he delivered annual courses. Like many earlier and later eighteenth-century popular science demonstrators, he toured the provinces during the early 1780s, in his case the Midlands and the North. To transform abstractions into concrete knowledge for an upwardly mobile audience, this public educator drew upon his early training as a weaver. Thus he envisioned the Newtonian principles of attraction and repulsion as an effluvial tapestry of intersecting powers materialized as tangible points, spheres, and rays of light (fig. 135). Palpable metaphors of vat, cloth, and thread also colored the descriptions of his electrical performances. When a high-flying kite caused a sensation "like a cobweb" almost biblically to pass over his face, Walker was certain he had tapped into the "cosmic Leyden Jar" (see fig. 131, no. 32).[92]

Body experiments, on one hand, possessed a physicality and directness remote from costly, complicated, and difficult-to-handle precision instruments such as Lavoisier's large beam balance, calorimeter, and gasometer.[93] Significantly, David's monumental double portrait of the great French quantifier and his artist-wife emphasized Lavoisier's private possession of

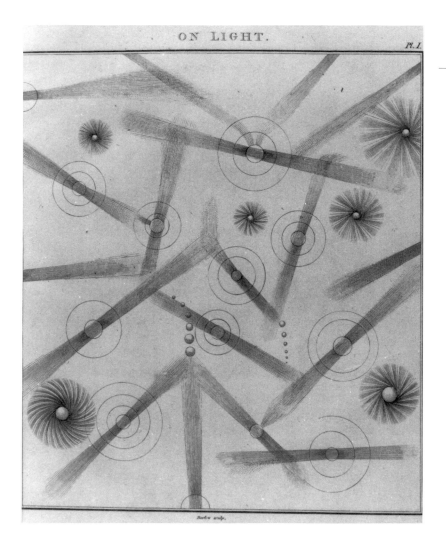

ON LIGHT.

135

Adam Walker

*On Light*

1799

from *A System of Familiar Philosophy*

theoretical knowledge. He is shown engaged in writing, in the construction of a rationalized nomenclature, not in the act of observation or the
public demonstration of material and experimental data (see fig. 95). On
the other hand, the lowbrow antics of "montreurs de curiosités" resembled
the voyeuristic illustrations to an eighteenth-century *roman noir.* We have
seen how the most intimate somatic spaces were aggressively exploited or
violated in the cause of making education both entertaining and
profitable.[94] Male and female viewers alike cringed in fear at the sight of
raw energy registered as horrific torment. Indeed, there was something
seductive and disturbingly "feminine" in the experimentalist's partial
nudity and the relinquishing of identity to the invasion of destructive
forces outside, and greater, than himself.

At the same time, the science demonstrator was literally the instantiation of the artist as controlling *plastes* or *fictor*. Joseph-Aignan Sigaud de La Fond (1730–1810), son of a clockmaker, practiced obstetric surgery before turning to the study of inorganic phenomena. Significantly, his conversion occurred after he audited Nollet's Parisian lectures. In 1759, he became *démonstrateur de physique expérimentale* at the Collège Louis-Le-Grand and, the following year, succeeded to Nollet's chair, supplementing the latter's lectures with courses on anatomy and physiology.[95] His complex and varied career interests us only insofar as it, too, belonged to the ambiguous climate of late eighteenth-century popular instruction based as much on conspicuous self-transformation as on shaping the public. Along with other celebrated electricians of his generation, Sigaud was a sculptor of the self, possessing the *terribilità* of the creator and destroyer who vanquished his medium. Instead of Michelangelo's intractable marble block, it was his own flesh that was mortified by Satanic expertise and shudder-provoking instruments.[96] Beautiful, well-crafted machines of his devising made ideas visible. They exhibited the formative powers of art to overcome external obstacles while incarnating the experimentalist's internal struggle to express himself through ingenious manipulations (see fig. 96).[97]

The aesthetics of eighteenth-century technology, then, was playful, sensuous, artful, mannered (derived from *manus*, hand).[98] Originality fostered experimentation and encouraged a physiological behaviorism that went beyond rational measure or calculation. In the end, the operator's organism became a grotesque hybrid of feeling tissue and inscribing apparatus. A teaching toy in new guise, the virtuoso's mechanized body inverted Vaucanson's sensualized androids. Finding the limits of human physical performance inspired both the experimentalist's athletic regimen and the inventor's myth of indefinite physiological perfectibility.

### Automata as Learning Machines

The shadowy world I have been describing composed of mathematical recreationists, "improving" conjurors, empirical artists, popular science demonstrators, instrument manufacturers, and entrepreneurs was populated by a group of men attuned to signals from the market. Resolutely on the side of the practical moderns, they belonged to an industrial sphere falling outside the rational orbit of the *Encylopédie* and the theoretical pretensions of professional academies.[99] Guyot and Hooper, Decremps and

Halle, Greuze and Wright of Derby, Nollet and Martin, Duchesne and Newbery were allied in the belief that cognitive mastery is a form of making. Dedicated to the principle that art and science were experimental, they sought ways to make them function in the public realm.

Living examples of a new man-machine relationship, these hands-on practitioners only comprehended when they constructed. They were thus descendants of that dynamic Baconianism which Antonio Pérez-Ramos has called "the maker's tradition."[100] Combining the savant, worker, and showman into one, these designers of simultaneously cognitive and material artifacts, intelligible and enjoyable goods, offered an epistemological alternative to the mathematization of the physical sciences.[101] The training of technical men, the realization of works, and the appreciation of the applied sciences emerged from the engagement with an artisanal world. Yet ingenious devices capable of reproducing nature's effects were more than utilitarian utensils. They were tangible conceptions and thus the engaging eighteenth-century equivalent of today's interactive and eerily realistic computer programs.[102]

The clockwork routines of the science demonstrator found their counterpart in the quasi-human intelligence of the automaton. The flesh and blood *homme-machine* was the dangerous double of the corporealized contrivance that acted as if matter could feel and think.[103] Jacques de Vaucanson's (1709–1782) famous cybernetic constructions were intimately connected to his lifelong search for the perfect artificial person (fig. 136). Eager to investigate the mechanical elements in man himself, the great inventor's feats of dexterity recalled the precision and agility of jugglers, acrobats, and magicians. His inquiry into the relationship between mental and physical events remains a major interest today among philosophers, cognitive psychologists, neuroscientists, ethologists, and artificial intelligence researchers.[104] This brilliant engineer, who invented the entrancing *Flute Player, Provençal Shepherd,* and *Duck,* also devised the automatic crank that revolutionized weaving at the turn of the nineteenth century. As official inspector for France's silk industry, he, too, inhabited this ambiguous region of material production. Crafty labor practices, as we have seen, were murkily intertwisted with superstitious folklore, tricky finance, popular culture, gimmicky apparatus, recreating charlatanism, and capitalist education.

Jacques de Vaucanson

*Automata (le flûteur automate,*

*le canard artificiel, et le provençal)*

1738

from *Le Mécanisme*

H. Gravelot delin.

Vivares Sculp.

Brisson's *Dictionnaire* defined the automaton as "a mechanical piece acti-vated by springs, or weights, or some other force whatever it might be, & which contains within itself the principle of its own motion."[105] In 1738, Vaucanson published to European acclaim a self-promoting account of his first stunning creation. While musical clocks, rotating spheres, and mechanical orchestras were common during the eighteenth century, the life-size *Flute Player* was singular in not merely imitating the sounds of instruments but in simulating the biological functions responsible for the *tour de force* artificial effects. Vaucanson's publication laid stress on the variable pressure of fingers and the separate beats of the tongue required to perform eleven different airs. In his domestic laboratory at the Hôtel de Longueville, with its sales and exhibition gallery set up in the Salon des Quatre Saisons, the industrialist went to extraordinary lengths to manufacture lifelikeness. Going so far as to upholster the right arm and hand of the flutist with real skin, he not only concealed the chains but attained the softness of living tissue to produce the nuanced effect of touch during a live performance.

This wily designer, perpetually at the mercy of creditors, tantalizingly described maneuvers that he did not illustrate. Thus the narrative informed the reader that the pursing of lips and rise and fall of the jaw entailed four different devices. Eight levers were responsible for modulating the intake of breath controlled by the mouth. The potential consumer was seduced by the invisible "miracle" of over two hundred bellows and levers that somehow worked together to produce a single natural impression. Adumbrating Heinrich von Kleist's inhuman marionettes, elegant and sensual automata possessed an analogous hidden source of superhuman grace. One thinks, too, of Mary Shelley's Victor Frankenstein who dreamed of chemically generating a mechanical man. Although lacking eighteenth-century charm, the monster was similarly the product of technological *hubris*. Vaucanson proudly claimed to imitate "by art everything human kind is compelled to make."[106]

Such enormous erudition and virtuosity were tellingly compromised in an ambitious scheme to offer experimental verification of an important and contested physiological function. Vaucanson was a partisan of Hecquet's by then outmoded hydraulic model of the digestive system. One of his automata became an instrument in a public clinical trial involving the audacious challenge to Boerhaave's and Helvétius's "méchanisme des viscères," founded on the fashionable principles of trituration.[107] The notorious gilded copper eating, digesting, and defecating *Duck* was minutely copied after nature. It quacked, splashed around in the water, and drank. Some four hundred pieces were needed to imitate the flapping wings, while myriad gears and pulleys, concealed in the pedestal, mimicked the distinctive way in which it lifted its feet and swiveled its head. The most astonishing process, however, was hidden from sight: the iatromechanical dissolution of grain in the stomach and the transportation of this pulp by tubes to the anus where it was evacuated through a mechanical sphincter.[108]

Having witnessed the *Duck* in action on one of its many tours, the Berlin enlightener Christian Friedrich Nicolai revealed, in 1783, that the supposed chilifaction laboratory in the intestines was a hoax. Seeds actually only entered the breathing tube and not the stomach. Apparently Vaucanson discovered to his dismay that the pulverization of the nutrients took too long during fermentation. A well-known proponent of the three humor theory, he felt obliged to prepare a fluid mass resembling solids sieved by saliva, pancreatic juice, and bile. Deceptively deposited in the bird's posterior, feces could be expelled at the desired moment by a triggering mechanism. Pointing toward the late twentieth-century concern with academic fraud and scientific integrity, it is noteworthy that the experimentalist tried to cover up his ruse. While the *Flute Player* and the *Provençal* could be examined at length during their European travels, the *Duck* was available only for the briefest inspection.[109]

Not coincidentally, then, the aims of instructive science, amusement, and the marketplace were uncomfortably mingled. These popular demonstrations of automation were also intended to draw attention to Vaucanson as a virtuoso crafter of purchasable commodities. The master's larger project of creating an artificial man was taken up by Pierre-Jacques Droz and his son, Henri-Louis. Intricate wheelworks, rotating cams, and levers set in motion a trio of famous automata by them, still on display at the Museum of Art and History at Neuchâtel. In the early 1770s they created the life-size *Scribe,* a three- or four-year-old barefoot boy seated at a mahogany desk who wields a goose quill that he dips into an inkwell. With perfect penmanship, the pupil somewhat jerkily composes a short text, all the while following the scratchy motion of his hand with his eyes. The diligent writer's companion is the Chardinesque red-jacketed *Draftsman.* This young lad flourishes chalk, and, with the aid of bellows concealed in his head, occasionally stops to blow dust from his paper (see fig. 3).[110] The central figure of the mechanical group is the blond *Musician,* a teenage girl in an elegant blue and gold dress who plays the clavecin and bows to her audience at the end of the performance. Moving her head and eyes, she visibly inhales as her chest rises and falls. Such simultaneously uncanny and enchanting displays of the clockwork human at work implicitly and explicitly challenged the manual trades.

Like today's learning machines, "moving anatomies" and artful "androides" were prized technological property fueling the cult of originality accompanying a booming eighteenth-century information industry. Extraordinary inventions were valued within a competitive market for beautiful and diverting cultural products in which it was often difficult to separate scientists from schemers, artists from copyists, poets from plagiarists.[111] Significantly, Senebier praised Vaucanson, Réaumur, and Duhamel de Monceau as practically oriented experts complementing the rarified atmosphere reigning in the Académie des Sciences and the Institut National. Yet these applied engineers and unreflecting technologists, he argued, groped by trial and error. The darker side of this material culture of tinkering, with its encouragement of personal vanity and pretense, called for skepticism and the moral and intellectual *dirigisme* of the critic. Perpetuating a longstanding dualism between mental and manual activity, Senebier claimed that the rational observer must guide the servile craftsman just as the dexterous surgeon aspired in vain to the theoretical knowledge of the anatomist.[112]

Vaucanson's dynamic figures belonged to the Enlightenment ethos of unlimited progress. Smoothly performing machines, masquerading as human beings or animals, seemed to hold out the promise that organisms could become infinitely perfectible by blending muscle with metal. Sheer record-breaking energy became a virtue. The eighteenth-century frenetic automaton was thus a profound symbol of modernity. It looked ahead to Marinetti's speeding racer as well as to the technologically manipulated body of the contemporary high-performance athlete.[113]

### Digital Performances

Vaucanson's sleight-of-hand in the name of experiment proved an embarrassment to the Age of Reason. Automation disconcertingly turned human beings into things and conferred a dubious existence on material objects. Moreover, as the acme of simultaneously painful and pleasurable manipulation, it troublingly alluded to a bygone finger-using, or digital, era that quantifying scientists were anxious to forget. Battling the growing trend toward abstraction, Rousseau and Court de Gebelin reminded their readers that pantomimic speech, or language's earliest and performative incarnation, was mnemonic and concrete, literally picturing the mind's grasp of concepts (see fig. 1). Like palpable charms, amulets, or fetishes, expressive gestures did not just reflect thought, they helped constitute it by making elusive ideas physical and visible. We noted in chapter 1 how the later eighteenth century was the birthplace of modern mass literacy drives. These fostered the displacement of Orphic oral per-

Adam Walker

*Mechanics*

1799

from *A System of Familiar Philosophy*

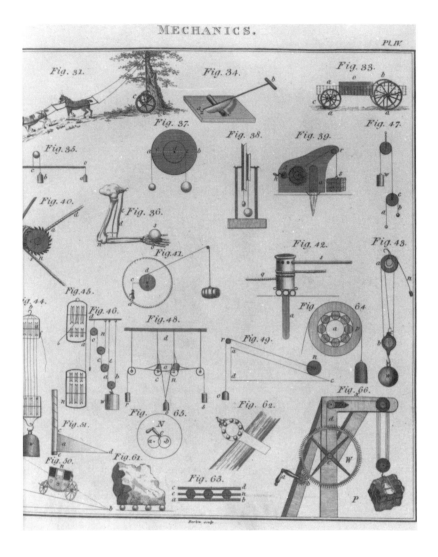

formances by silently read texts. This "primitive" or representational dimension of communication—composed of dramatically enunciated iconic, metaphoric, and symbolic messages—became increasingly undermined through disembodied data cast in lead.

In the first instance, then, the automaton signified the height of repetitive adroitness. In the second instance, it stood for the seductive evasion of work, for the voluptuary ghost in the machine who mysteriously, effortlessly, but, above all, handlessly, usurped human initiative. Not surprisingly, the growing suspicion of the magical "machinist" who crafted luxury products to cure the *ennui* of jaded purchasers has deep roots in the agrarian economy of the old regime. If it can be said, as Rémy Saisselin argues, that the baroque is an extended civilization that died with the extinction of a certain kind of society, then the Enlightenment represented a pan-European intellectual movement criticizing the conspicuous spending and hedonistic vices associated with an urbanized mercantilism.[114] Before the Industrial Revolution, the extravagance flaunted by *nouveau-riche* merchants stood in stark opposition to the simple hand-driven tools and farm equipment associated with the frugality of traditional life on the land (fig. 137). In this sense, Vaucanson's virtuosity, ostentatiously displayed in expensive artifacts, provided a counterimage to virtuous rural labor. His clever handling epitomized those provocative and fashionable goods beguiling big-city dwellers. Hence the double face of modern technology: as the corrupting production of nonessential wares and as the guarantor of leisure by easing drudging tasks. This conundrum served to further embroil questions of commerce in aesthetic and moral disputations.[115]

Bruno Latour's claim concerning the irreducibility of the sciences[116] strangely omits a discussion of what might be termed ordinary marvels. With the aim of giving the aesthetics of everyday economics its due, I want to situate the varieties of handwork among the daily and prosaically commercial transactions in which it took shape. Forgotten associations quickly emerge. The mathematical recreationist's teaching tools, the juggler's brass balls, the experimentalist's electrical apparatus, the inventor's showy instruments (fig. 138), the industrial designer's costly products (fig. 139), all smacked of expenditure and sophistication. They seemed the "unnatural" opposite to the simplicity of the preindustrial setting, with its blazing smithies, smoking furnaces, and iron forges resonating to the beat of hammers on anvils (fig. 140). Compared to a metropolitan economy fueled by artifice, such "natural" sites and implements connoted honest physical toil by the humble as ordained by God.[117] To complete our exploration of the negative perceptions eddying around the legerdemain practices of the crafts and trades, we must look at the materialities of manual expression in the conduct of practical affairs (fig. 141).

For cultural primitivists, the problem with manufactured commodities was not unlike that dogging intellectual acquirements. Rousseau's conviction that too much importance was given to writing derived from the belief that letters had become merely mechanical and demystified counters within an abstract system of exchange. Accordingly, he admonished tutors not to give children books. Emile was thrust into the artisan's workshop,

138

Jacob Leupold

*Compass, Protractor, and Conveyer*

*(Instruments for Trigonometry)*

1727

from *Theatrum Arithmetico-Geometricum*

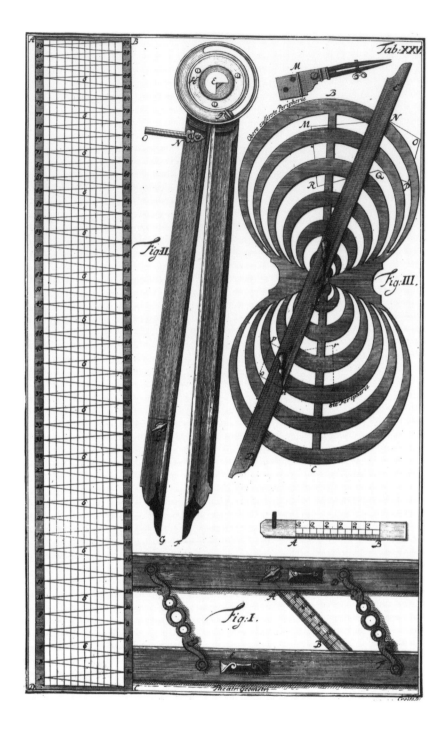

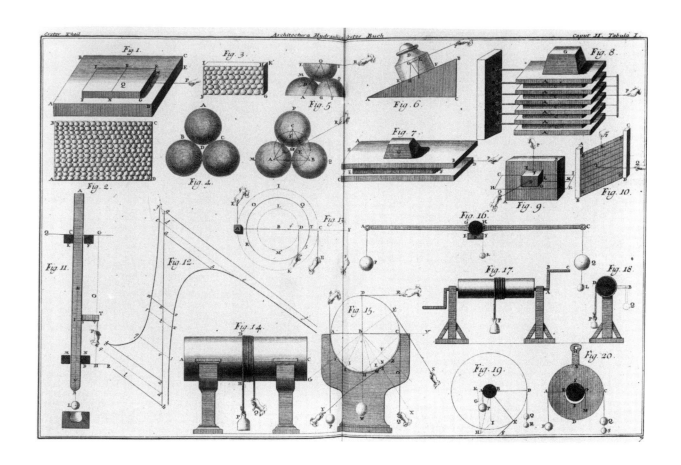

139

Bernard Forest de Bélidor

*Pulley and Weights*

1740

from *Architectura Hydraulica*

140

Joseph Wright of Derby

*The Blacksmith's Shop*

1771

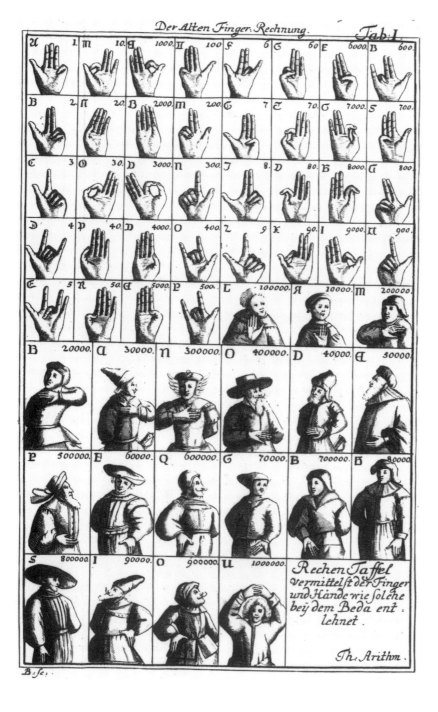

Jacob Leupold

*Demonstration of Dactylology*

1727

from *Theatrum Arithmetico-Geometricum*

in contravention of current pedagogical trends, where he learned to use his hands.[118] These exercises were intended to defy automatic writing, or an undue privileging of nonmimetic language. Romantically going against the grain of learning solely conceived in terms of literacy, this passionate educator foresaw the movement from Kant to Wittgenstein that claimed linguistic analysis was the only worthy task of humanistic education.[119] The materialism of vivid speech acts and the tangible exertions of elocution were being challenged, then and now, by attempting to reduce knowing to the immaterial signifier.[120]

This tension between automation and handwork, abstract language and physical expression, conception and technique, also characterized apparatus. There were two sorts of instruments: unnatural and natural. Technology could be rarefied or sensuous,[121] taking the form of closed-system calculating machines (fig. 142) or of phenomenal devices that hieroglyphically pictured the perceivable properties of air, earth, water, and fire (see figs. 96, 99, 111). *Appareil,* signifying joining together "a collection of machines or instruments necessary to conduct a series of experiments on a determined subject," was symptomatic of a new mathematical rigor associated with elaborate and difficult-to-handle equipment.[122] These ancestors to present-day Big Science combines, headed by "macro-actors" crushing a multitude of laboring observers, again caused Rousseau to fulminate. "This whole apparatus [*appareil*] of instrument and machines displeases me. The scientific atmosphere kills science. Either all these machines frighten a child or their appearance divides and steals the attention he ought to pay to their effects." The Genevan's exhortation to the experimentalist to make his own instruments was anachronistic. The ability to find meaning in forms, colors, odors, and textures tended to disappear in the post-Lavoisier era of the 1780s, enamored of quantifying setups.[123]

Recall how simple procedures and acute sensory awareness has turned the body of the popular science demonstrator into a teaching tool. This living example was being gradually superseded by the professional's automated machinery. Just as Rousseau's *sensible* daydreamer was threatened by the ideal of a mechanized person, the error-prone hand was under attack by the calculator.[124] Rousseau's diatribe against standardized and sanitized inscription devices was a folkloric defense of the endangered laboratory as a material environment accessible to all, and of manual dexterity as a necessary accomplishment for every many, woman, and child.

Joh. Poleni
Rechen Machine

Jacob Leupold's (1674–1727) *Theatrum Arithmetico-Geometricum* (1727) brilliantly captured that ancient and virtuosic visual-kinesthetic communication that was being dragged into the post-Cartesian realm of abstraction. Visual counting, like the conjuror's free shows, was part of a somatically conveyed sales pitch.[125] In business transactions, the body as spectacle similarly became localized in the choreography of the hands. Like the *scarlatino*'s legerdemain tricks, legitimate mercantile routines thrived within a commercial setting.

Leupold, a Czech engineer, documented how numbers continued to exhibit their primeval and nonabstract beginnings as body parts within the oral-visual culture of the late baroque (fig. 143). His northern origin is significant because it situates him within a society that valued *handling,* the virtue and virtuosity of fastidious, neat, and "cunning" manual execution.[126] A historian as well as a machinist, Leupold conjectured that Roman numerals originated in digital imagery. He demonstrated how three-dimensional patterns, made by flexing the fingers, literally incarnated ciphers and characters. Such personified mathematics can still be discerned in the initial letters spelling the Latin names for the appropriate figures. According to this tangible hieroglyphics, the C-curved palm represented *centum.* Thumb and index formed into an L stood for 50. D equaled one-half of *mille,* or M, and so on.[127] The pictorial system of assigning arithmetical values to letters fueled the mystical combinations, artful calculations, and divinatory speculations of hermeticism and the cabbala.[128]

Trained in theology at Wittenberg before turning to mathematics, Leupold knew that Appian, the Venerable Bede, and Aventinus had been fascinated by *manuloquio,* or natural "language with the hands" (fig. 144). He thus linked counting to a global and synthetic medium of prearranged gestures transcending arithmetic to embrace the infinite nuances of human expression: crying, pondering, doubting, grieving, suffering. Dactylology, or "the art of communicating ideas by signs made by the fingers," belonged to those orchestrated operations constituting the universe of nonverbal behavior conveyed imagistically during a performance.

The engravings to Leupold's magisterial work show both Roman and Hindu-Arabic notations. Each practice had long traditions of expertise and different social overtones.[129] The former, with its classical connotations of an intractable linear geometry, presupposed manipulating tactile

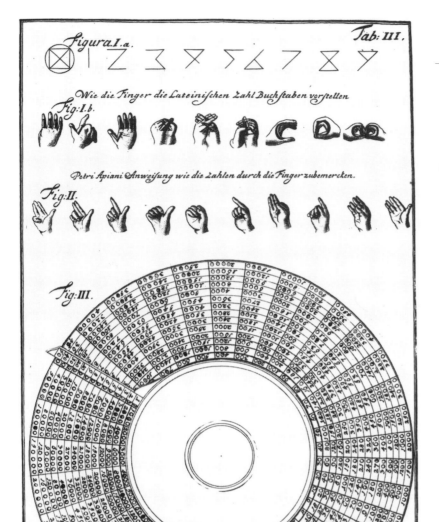

143

Jacob Leupold

*The Connection between*

*Roman Numerals and Letters*

1727

from *Theatrum Arithmetico-Geometricum*

144

Jacob Leupold

*Manuloquio, or Language with Hands*

1727

from *Theatrum Arithmetico-Geometricum*

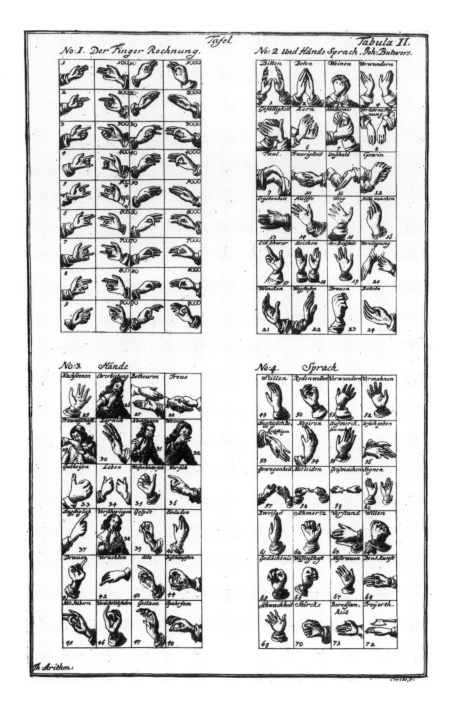

beads arranged in rows on an abacus. The latter, akin to the speedy cursive favored by businessmen, resembled the airy or balletic beating out of units and tens on the left hand and hundreds and thousands on the right (see fig. 141). From this algebraic *ars combinatoria,* connected to the vernacular rather than to Latin, emerged the shifting fluidities of the romantic arabesque. Whether considering the chemical idiom of Novalis, the pantheistic hieroglyphics of Runge and Friedrich (see fig. 16),[130] or the timeless sign language of Blake, creation consisted in revelation not in automation (fig. 145).[131] The instrumental and utilitarian gesticulations of Leupold's transacting journeymen—identical to the corporealized alphabets of deaf-mutes—were poetically transformed in the British illuminator's image of *Christ.* As the epitome of the liberal artist, Jesus' ritualized hands were metamorphosed into eternal and pure symbols free of the taint of lifeless mechanism.[132]

Conversely, the raw physicality of the speech act gestures documented by Leupold contributed to the carnality of rococo painting. As graphic means for inserting obscene material implicitly into an utterance, they managed to circumvent common verbal taboos. Greuze, in particular, was a master of emblematic and equivocal positions. His work is a visual encyclopedia of classifiable optical "statements" ranging from undisguised Italianate insults to the quasi-pornographic, but elliptical, actions of his ring- or orifice-indicating maidservants (fig. 146). Components embedded within the larger social code of gallantry, these unspoken words were violations of *honnêté* and existed only during a visual performance. The vulgar forearm jerk, dismissive flathand flick, sly cheek stroke, disdainful nose thumb, and roguish ear or mouth touch (fig. 147) were popular emblems possessing great historical longevity.[133] It is not coincidental that they flourished within the manual ambient of the baroque and rococo periods, still attuned to the control of others' behavior through virtuosic manipulation (fig. 148). Blake's epiphanic *Christ,* simultaneously blessing and revealing, thus resembles La Tour's sardonic *Self-Portrait,* mockingly directing the viewer to enter its *trompe-l'oeil* space, as well as Greuze's arch genre scenes. In each case, the very different sets of hands express a preconceived meaning.

One of the ways in which romantic painters redirected these eighteenth-century engineerings of gesture can be seen in their shift in attitude toward movement. The displacement from naturally connective to artificially disjunctive motions not only altered the material bond of the speaker to his or her members but implied a realignment of the relationship between the mechanical and the organic. Limbs were represented functioning like instruments, literally turning sitters into technological men. Goya's android, *Don Manuel Osorio Manrique de Zuñiga,* is an eerie construction precisely because of the robotic manipulations he distantly performs on his pet magpie governed by a lead (fig. 149). Acting as if by remote control, the arms appear detached from the boy's inner consciousness as well as his physiology.

It is equally unclear what anatomical or psychological links exist between Franz Pforr's stony fingers—carefully poised on the window ledge of a vine-overgrown Gothic porch—and his faraway gaze. The Nazarene painter is transmuted into an inorganic specimen fantastically framed within the museum case of an idealized medieval world (fig. 150). Preternaturally still and heroically scaled, he almost obliterates a diminutive

146

Jean-Baptiste Greuze

*The Laundress*

1761

147

Jean-Baptiste Greuze

*Espièglerie {Roguish Trick}*

c. 1782

148

Maurice-Quentin de La Tour

*Self-Portrait*

c. 1737

149

Francisco de Goya

_Don Manuel Osorio Manrique de Zuñiga_

1784

150

Johann Friedrich Overbeck

*Portrait of the Painter Franz Pforr (1788–1812)*

1810

151

Jean-Auguste-Dominique Ingres

*Studies for "The Martyrdom of Saint Symphorien"*

1833

virginal figure. Her busily clicking knitting needles form an incongruous counterpart to Pforr's masklike features, unbending torso, and quiet hands. Male thoughtfulness is thus contrasted to feminine doing without knowing. Her absorption in merely mechanical activity is allusive of the disruptive and mindless restlessness that had been associated with rococo craft.[134]

In Goya's psychological study of the personified fears encircling childhood or Overbeck's symbolic portrait of his young friend, doomed to drown in the Tiber, the sitter's hands no longer behaved as conventional emblems. Rather, they functioned as precision tools capable of exhibiting the discord between sensation and ratiocination. This greater abstraction, I believe, was congruent with a mentalized outlook that valued "thinking" apparatus moving intelligently on its own, not brute materials requiring tradesmanly manipulation. Analogously to the later eighteenth-century drive toward "objective" calculating machines (see fig. 142) and self-propelling automata, painters conceived gestures as free agents operating like independent units (fig. 151).

In all facets of practical life, the artisanal play of fingers had been integrated, with varying degrees of success, into the coherent structure of the body during the baroque and rococo eras. Even broken speech and interrupted conversation might be fully enacted in another medium. This gapped information, requiring viewer participation to complete the idea, would turn into fragmented transmission in Ingres's art. Disembodied gesticulations, uneasily collected from diverse historical sources and a treasury of European styles, now exhibited only incoherence. The French painter's spectacle of polarized elements was symptomatic of the nineteenth century's struggle with the increasing division between controlling technology and wayward biological substance. The represented body had become a kaleidoscopic gallery of conflicting and paratactically juxtaposed curiosities. Each segment was a singular example crying out for special arrangement and constant rearrangement. The installation of choice forms into a composition was never final since the decision came from the outside, not from within a vital organism.

EXHIBITIONISM

## Spectacle of Nature

To show, display, or exhibit has become an obscenity among deconstructivist and poststructuralist critics.[1] Reminiscent of Vaucanson's *Duck,* mass media stands accused of eating, digesting, and excreting world culture.[2] Television's assimilation of diversity and homogenization of the alien is based on the ability of immaterial images, unlike national languages, to travel passportless across borders. Penetrating everywhere and feeding on one another, they risk becoming fetishes, ambiguous and contextless commodities for consumption.[3] The art and entertainment format, especially, is open to the charge of sugarcoating education by purveying images requiring no conscious effort on the part of the beholder.[4] Hiding the mechanisms behind visual construction is like window-shopping. No longer seeing the constitutive technology encourages the ingestion of a seamless spectacle of goods.

No contemporary event better illustrates our involuntary linkage to a global network of images and the existence of a universal imagism than Benetton's recent advertising campaign. Not content to parade its merchandise, the Milan-based knitwear company engages in sleight-of-hand. Striking and often unsettling photographs bought from newspaper archives are published "unimproved" in magazines and posted on billboards. A young man dying of AIDS, an infant with umbilical cord still attached, child labor in underdeveloped countries, or refugees fleeing from disaster are used to merchandise fashion. While not unique in using political causes to shape a multinational reputation, Benetton is of interest because its layouts appear without change in 103 countries and are wordless except for the presence of the corporate logo.[5] Just what is being communicated by these pictures: ethnic awareness and environmental concern, guilt-free consumption, or the alleviation of buying anxiety? My point is that it is difficult to say.

Having turned images into products and products into socially constructed images, the Benetton campaign raises a larger question concerning the ethics and aesthetics of showing and seeing objects. This is the identical problem today's museums face as they wrestle with the "poetics and politics of representation."[6] Significantly, the current dilemma is often cast as a dualism: are artifacts worthy of display because of their moral and social significance, or because of their formal and timeless properties? I suggest that this dichotomization, pitting worthwhile observation against empty voyeurism, received its distinctively postmodern twist in the eighteenth century.[7] Recall how showmen gave experimental science a bad name. False projectors were accused of producing only stupid admiration in bored gentry.[8]

This chapter, then, explores from another vantage the tension between an Enlightenment classifying culture, or rational systematics, and a waning baroque oral-visual polymathy. The undemanding "spectacle of nature"— encouraging an effortless gaping among the leisured classes—was challenged in the second half of the century by new professionals forging exacting taxonomies. Current focus on the rational order shaping Enlightenment "museum culture"[9] has served to obscure, however, the existence of meaningful *Wunderkammer* arrangements prior to the advent of scientifically sequenced works. Contaminated by crowd-pleasing spectacle, these collections resembled the razzmatazz natural philosophy taught by itinerant lecturers. Hence the cabinet of curiosities, too, existed ambiguously in between entertainment, performance, and practical instruction. Browsing nature for possible possessions was akin to shopping. Infinity could be made manageable in consumable chunks.

My purpose is not to rewrite the history of museums in terms of what one author has called the transformation from temple of beauty to cultural fair.[10] Nor do I want to rank them, as recent critics have, among those corporate "public spaces" such as amusement parks and malls.[11] The intent throughout this book has been to search for the positive roles played by images, past and present.[12] So it is to the notion of the exhibit as a school for instruction through perceptual patterning that we must now turn (fig. 152). Giovanni Paolo Pannini's (1691–1765) *Modern Rome,* a pendant to the celebrated architectural and landscape painter's medley of *Ancient Rome,* exhibits the ostentatiously accumulative character of pre-Enlightenment art and science. Significantly, in the discussions over early modern "museum culture," such a fantastic spectacle of heaped-up paintings has not been linked to the "curious and eclectic" mentality of the forgotten baroque and rococo *polyhistor.*

Collections, as part of the economic and social history of distant goods mobilized to ship home, required a human center. The virtuosi seated in Pannini's imaginary temple of art could act at a distance on many far-flung regions. Either personally, while on the Grand Tour, or more remotely through the intermediary of foreign agents, wealthy men of taste exercised

152

Giovanni Paolo Pannini

*Modern Rome*

1757

power through the expenditure of money.[13] What interests us is how the artificial and natural objects they brought back to such European capitals of knowledge as London, Paris, Berlin, or Vienna were recombined and made meaningful in new contexts.

Polyhistory, as a method for collecting and shaping information, can be traced to the *copia rerum ac verborum* of Roman grammarians and rhetors.[14] The oral-visual ideal of copiousness, easily straying into license or superfluity and thus offensive to later rationalist critics, reached its apogee during the "mannered" sixteenth and "Jesuitical" seventeenth centuries. The tension between an aggregate of information, predicated on an optic-based epistemology permitting surface connections to be made across many fields, and deep systems analysis became acute.[15] First attacked as charlatanism by a Cartesian subject-oriented skepticism, the aspiration to gather together everything was finally demolished by the high enlighteners. D'Alembert, the internationally renowned mathematician and coeditor with Diderot of the *Encyclopédie,* derided polymathic hodgepodges. In the preliminary discourse, quantified physics emerged as the basic science to which nature had to submit. Diderot, on the other hand, who labored on the Herculean project of producing 17 volumes in 21 years (1751–1772), was influenced by polymathy's materialism, its concrete habit of putting all sensory and intellectual domains in imaginative and dialogical communication with one another. His love of instruments and celebration of technical processes also belonged to the polymathic universe of crafty *ingenium* embodied in the applied arts, marvelous machines, and skilled trades.[16] For the rationalist d'Alembert, however, tesselated *variora* were unsystematic because they were founded on false analogies.

Yet, like Diderot's conversational notion of the *Encyclopédie* as an extension of the *salon,* the old compendious learning emphasized the free commerce between disciplines. Consequently its fluid combinatory style has much to offer those late twentieth-century scholars dissatisfied with the "imperialism" of a "master" text or the ossification of traditional disciplinarity.[17] Moreover, the experimentalism of polymathy provides an ad hoc model not only for the rejection of essentialism but also for avoiding those distortions of non-European perspectives possible within a program of European studies.[18]

Like eighteenth-century popular science demonstrators, polyhistorians hunted after unpredictable connections and jumped across scholarly borders then being erected and subsequently codified by *philologi* or *critici*.[19] Important, too, was their presentation of information as miscellaneous spectacle as opposed to the nineteenth century's higher criticism with its anatomization of dead sources into lifeless bits of antiquarianism. This tension between aggregate and system can be reformulated as a battle against the disruptive element of the figural. Jean-François Lyotard has argued that the aesthetic is the transgressive realm where libidinal desires remain uncontained and so reveals the limitations of an ordering theory.[20] Yet his view that designs are relatively free of meaning because they are not determined by philosophical traditions or linguistic structures is, in the end, deeply logocentric. This tenacious bias may be unseated only if patterns can be shown to display and say things texts cannot.

Historically, museums took the form of complex buildings, often organized around a domed or vaulted dominant hall flanked by wings decorated with *naturalia* or *artificialia* (see fig. 152). They also existed as domestic spaces set apart specifically for personal collections (fig. 153). These contrary sites mirrored a growing intellectual schism. Heroically scaled public monuments, designed for scientific classification and mental pursuits, became increasingly differentiated from private, recreating repositories of heterogeneous material possessions.[21] North of the Alps, as in the *Kunstkammer* depicted by Frans Francken II, the pleasures of visual experience superseded systematic arrangement. Indeed, the heteroclite items forming this artist's gallery presented a *theatrum* of *visibilia* to be looked at and enjoyed rather than to be fitted into a genealogical narrative.[22] As in the eighteenth-century amateur's *cabinet* (see fig. 43), books, prints, paintings, statues, gold and silver vessels, musical instruments, medals, shells, corals, and fossils demonstrated the material ingenuity of crafty artisans and the skill of sportive nature. Rarities were conjured into life by the activity of looking and so were related to Nollet's and Trembley's laboratories, well-stocked with tantalizing machines and odd specimens (figs. 41, 42, 47). As quintessential *visibilia,* they were similarly connected to Germain's or Chodowiecki's showroom-workshops (figs. 91, 97), crammed with costly samples, to Vaucanson's *salle d'exposition,* displaying automata (fig. 136), and to the magician's stage, littered with diverting apparatus (figs. 58, 72, 73).

Pansophia, identified with secluded or sociable Renaissance architectural units such as the *stanza, casa, casino, guardaroba, studiolo, tribuna,* and *galeria,* also characterized the Janus aspect of eighteenth-century muse-

153

Frans Francken II

_The Art Collection_

1636

ums.[23] Part alchemical chamber for the covert performance of hermetic rites (fig. 62) and part memorial to overt didacticism, the museum was pulled between being a grottolike hideaway, tucked in a scholar's private quarters (fig. 154), and a utilitarian and gregarious institution, like the public library. The analogies of these retreats and sanctuaries of the muses to the treasury, microcosm, and theater were founded on their shared reliance on spectacle.

155

Salomon Kleiner

*Rear Section of Open Display Cabinet*

1751

from *Christophori de Pauli Pharmacopoei Camera Materialium*

But it was the polymathic collector's three-dimensional encyclopedia of curiosities that catapulted museums into the realm of "hypervisibility" (fig. 155). The postmodern notion of a voracious gaze was first unleashed in the galleries, menageries, fairs, puppet shows, laboratory lessons, and indoor or outdoor amusements proliferating during the early modern period. These recreations gave rise to what the phantasmagorist Robertson termed "le genre si nouveau d'exhibition."[24] From balloon launchings (fig. 21) to masquerades (fig. 23) to the projection of apparitions (figs. 48, 50), "le spectacle nouveau" impressed itself on the restless and pleasure-loving imagination of the public.[25] Before examining in detail a specific eighteenth-century instantiation of visual polymathy as stage of marvels and spectacle of rarities, we need to ask how nature developed into a browsing field of pleasing fragments to gather, discuss, and gape at.

To get the most out of their aristocratic charges, tutors in Jesuit colleges thought it best to untether them regularly for about two hours a day. During the winter, *pensionnaires* turned the *salle d'étude* into a *salle de récréation,* forming a marked contrast to the dangerous idleness of the *externes* who lived in town outside the dormitory's protective walls (see fig. 71). Edifying card tricks and other harmless pastimes (see fig. 48) were replaced during the summer months by out-of-door games such as croquet, shuttlecock, and quoits.[26] The weekly half-day holiday and Sunday excursion into the country proved most significant, however, to the formation of a taste for natural history and the impetus to collect delightful portable specimens. This habit of the leisurely promenade, inculcated as part of the educational process, not only underlies Rousseau's sensation of reverie but the new eighteenth-century insistence on the health of enjoyable experiences unfolding in the open air.[27]

Importantly, children's and teachers' involvement in simultaneously pleasurable and instructive exercise was not limited to France. Empiricism and the observation of the natural world became a duty for all in post-Lockean England (see fig. 53). Paul Hunter has linked the rise of Protestant "meletics," or the occasional meditation on things encountered in even the most mundane circumstances, to the formation of a British "priesthood of observers and interpreters" increasingly supporting a science culture after 1691.[28] Further, the benevolent pleasures of sight, praised in the *Spectator* papers (first appearing in 1711), formed part of this veneration for the authors of the "new philosophy" and their magnificent theories of the earth and heavens. Both Addison and Steele were popularizers, wishing to bring Newtonian science into the orbit of gazettes and out of libraries. Like coffeehouse demonstrations and club experiments, the Scriblerians' daily essays were a highly visible form of public conversation, not private and pedantic writing.[29] As we have seen, this notion of spectacle as a way of demonstrating truth had been set in motion by Newton's *experimentum crucis,* involving the passing of a luminous beam through glass prisms (see fig. 115).[30] The theatrical staging of this phantasmagoric light show in repeated trials, and its memorialization in numerous popular accounts, lies at the root of an international fascination with nature's spectacle.

Like their rococo colleagues across the Channel, the male amateurs and virtuosi of the Augustan republic of letters valued information couched in the refined displays consonant with the labor-free ideals of civic humanism.[31] Both the French upper classes and the English gentry looked to the universe as a source of nontaxing learning achieved through contemplation. Horizontal skimming distinguished the *galant* or polite viewer from the vertical probing of the toiling professional. In this economy of discriminating gazes, however, women fared differently in each nation. Since the late seventeenth century, the female-organized and run *salon* had been the place where intelligent aristocratic ladies generated an informal and nonutilitarian type of instruction for themselves as well as for men.[32] While French women were able to engage in creative talk, their English counterparts were increasingly relegated to an amateurish copying after nature.[33] In the second half of the century, the denigration of craft by both British industrialists and high academicians facilitated the feminization, not of beholding but of the mechanical reproduction of views.

Unskilled laborers and peasants in the two countries generally lacked the means and the interest to enter secondary schools.[34] And, until much later in the century, they also lacked the English coffeehouse "penny universities" whose popular lectures were audited by printers, drapers, and weavers. Like the bored *externes* of the Jesuit *collège,* the largest portion of the populace, then, was "educated" through popular spectacles. They gaped at apparitions produced by magicians working the annual cycle of fairs and manning the optical cabinets spreading throughout Europe. A political print published in Nuremberg, *The Tea-Tax Tempest or the Anglo-American Revolution,* provides a rare glimpse into the cinematic interior of such an alternative space for education (fig. 156). Standing in for the carnival

The Tea-Tax-Tempest, or the Anglo American Revolution.
Ungewitter entstanden durch die Auflage auf den Thee in Amerika.
Orage causé par l'Impôt sur le Thé en Amerique.

156

Carl Guttenberg

*The Tea-Tax Tempest*

1778

conjuror, Father Time projects slides showing the destructive consequences of the British tax on tea to a horrified audience seated in a *camera obscura.* Rapt personifications of America, Europe, and Asia watch a phantasmagoric succession of scenes enacting the successful struggle against tyranny.

Since the seventeenth century, the task of the Jesuits had been to form Christian gentlemen by teaching them morality through religion and excerpts from the classics. Emphasis on good manners, polite talk, and the composition of essays devoted to noncontroversial topics made for pleasing breadth, not professional depth. In contrast, middle-class values—whether evinced in Chardin's attentiveness to how things were actually done (figs. 116, 122) or Wright of Derby's close, devotional observation of a suffocating bird (fig. 76) and smithies toiling over a hot blaze (fig. 140)—belonged to a demonstrative rhetoric at variance with aristocratic complacency (fig. 120, 124) and the taste for untaxing spectacle. Fontenelle's *Entretiens sur la pluralité des mondes* (1686) captured this contempt by people of high status, no matter their nationality, for comprehension gained through labor or commerce. In a series of six dialogues between an instructor and a lady of quality, the moral and philosophical implications of the new science were explored.

Comparing himself to Cicero, who was standardly read in epitomes at the *collèges,* Fontenelle also engaged in translation.[35] Instead of domesticating Greek philosophy for Roman readers, however, he converted the abstractions of the Newtonian system into delightful exchanges between "les gens du monde." The *secrétaire perpetuel* of the French Academy of Sciences declared: "I wish to treat philosophy in a manner that is not philosophical." Seeking to communicate science in an ingratiating style that would appeal to "tout le monde," he hoped even physicists might be diverted by the agreeable presentation of material known to them "more solidly."[36] Not only did this celebrated French author, himself the product of the *collège* at Rouen, choose the conversational mode (*entretiens*) and a woman as his pupil, but the lessons were set out-of-doors in the manner of the obligatory Jesuit Sunday outing. Looking at nature, therefore, was conceived as part of a sophisticated social configuration, not as solitary scientific observation (see fig. 114).

The "smiling" spectacle of the external world, as it emerged from the pen of an ardent defender of the moderns, resembled the seductive ambient

157

Jean-Marc Nattier

*The Princesse de Rohan*

1741

of an indoor assembly. Instead of being surrounded by the artifices of social encounter, Fontenelle's worldly onlooker was transported into a leafy *salon* (fig. 157). Jean-Marc Nattier's portrait of the *Princesse de Rohan,* depicted flipping the pages of the *Histoire universelle* in a green bower, seized such upper-class spectators of curiosities who needed their intellectual fare excerpted and sensualized. Aptly, the *princesse* was not represented pondering an abstract text. Rather, she was caught visualizing an already digested piece of information and then digressing from the pedantic volume to contemplate the attractive book of nature.

In his attempt to render science *galant,* Fontenelle drew an analogy between scenery and opera. Because of the angle, the spectator seated in the theater could not see the *machiniste* hidden in the wings or under the parterre. Scientists, however, had greater difficulties than set designers because the universe's wheels and pulleys remained concealed from them no matter the perspective. This antiillusionist stance emerged from the libertine tradition of late seventeenth-century French skepticism. Fontenelle's Cartesian epistemology of doubt led him to draw an important distinction that would come to a head in the second half of the eighteenth century. He pried apart a profound and specialized observation from superficial admiration. Like watching staged marvels or miracles, the beholding of bewitching phenomena was "a sort of magic," requiring nothing of the understanding.[37]

This separation between *la physique des savans* and *la physique des enfans* continued apace in Charles Rollin's (1661–1741) educational writings, important on both sides of the Channel from the 1720s onward. This Jansenist defender of the "Protestant" heresies of Port-Royal shared, nonetheless, with Jesuits the custom of conceiving the universe as a large picture in which every element had a use. Natural philosophy spread before the beholder "so beautiful a Spectacle," and taught him to observe order, symmetry, and the proportion of parts to the whole. The physics of the "learned," however, differed from that of the nonspecialist. The latter focused exclusively on objects making a direct impression on the senses, whereas the former searched for the invisible mechanisms behind appearances. The natural philosophy of children entailed "a study of nature which scarce requires anything besides the eyes, and for this reason falls within the capacity of all sorts of persons." Note that this is a fundamentally *aesthetic* approach based on "admiring" the "different beauties" of objects, "but without searching into their secret causes." Rollin, then, proposed an effortless theory of perception that was neither painful nor tedious, but a "recreation" and a "diversion."[38]

Ingesting the spectacle was helped along by epitomizing, by creating edited and compressed displays of nature's antiquities. Collections permitted the viewer to relive the perceptual moment of discovery by constantly forcing the eye to isolate and single out rarities from a controlled welter of competing phenomena. The analogy of natural history arrangements to artistic compositions led to the fashion of literally framing the bits and pieces of the physical world as if they were a remarkable picture or famous drawing.[39] Architectural surrounds encased the ornamental cabinets of private collectors (see fig. 155), while museumlike mountings accentuated individual specimens reproduced in luxury folio editions (fig. 158). This pictorialization was inseparable from conceiving nature's uniformity and variety as scenes for contemplation or observation. According to Joseph Priestley, "new fossils are perpetually pouring in upon him [the spectator]." For Adam Walker this optical deluge, deriving from the open "book of nature," spoke "an intelligible language to all understandings."[40] In each instance, easy apprehension was facilitated by subdividing the flow into watchable units permitting closer inspection.

The London physician and natural historian Martin Lister's sumptuous *Historia Conchyliorum* (1685–1692) represented a major early example of such lavish framing devices that also indicated a change in viewing distance. Reprinted at Oxford in 1770, the initial publication was the result of ten years of research and cost the then extraordinary sum of almost £2000 sterling. The 1,233 illustrations, distributed over 505 plates, were etched by Lister after drawings by his daughters Susanna and Anna. This compendium of strange stones supposedly exhibited the plastic powers of an animate nature. Making visible the invisible, relics from "deep time"[41] gave testimony of the earth's past activity. Opposing Nicolaus Steno's prescient argument on the organic origin of fossils, Lister (1638–1712) interpreted them as bizarrely patterned "sports" resulting from putrefaction.[42] Although a Fellow of the Royal Society, his *pictorial* theory of lithic hieroglyphics belonged to a baroque optical epistemology similar to that of Jesuits such as Athanasius Kircher.[43] Richly sculpted gilt borders and compositions calculated to accentuate randomness presented shells as picturesque sketches, not as textualizable documents of once-living marine animals. Occasionally, dynamic micromegalic juxtapositions could give way before static heroically sized natural monuments. Like those artful diluvian "singularities" examined in my *Voyage into Substance,* a colossal ammonite embodied the uniqueness of a freestanding natural monument.[44] Represented as a marginless masterpiece that had escaped the vicissitudes of time, it was shown literally without scale and beyond comparison (fig. 159).

158

Martin Lister

*De Musculis Fluviatilibus*

1770

from *Historiae Conchyliorum*

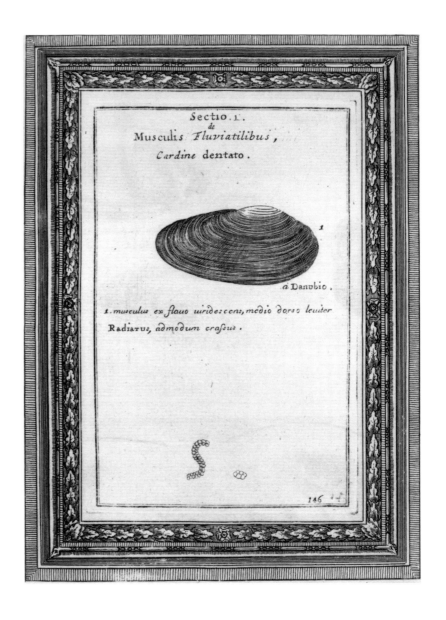

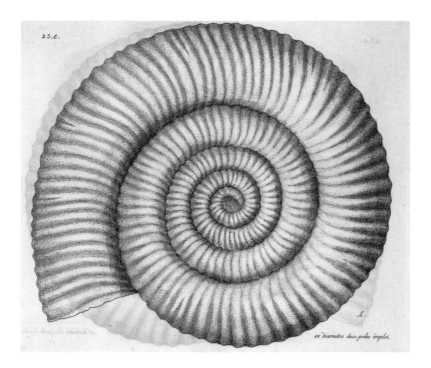

The "spectacle of nature" burst on the literary scene as a distinctive encyclopedic genre of visual stunners with Noël-Antoine La Pluche (1688–1761). Suspected of being a Jansenist, this professor of rhetoric was obliged to earn his living as a private tutor in Rouen on the recommendation of Rollin. In the early 1730s he moved to Paris, supporting himself by giving courses in history and geography. The wildly successful nine-volume illustrated *Spectacle de la nature* was sold out as soon as it appeared. Published between 1732 and 1750, it eventually went through 59 editions. Initially composed to serve as a science primer for his wealthy charges, it quickly became, according to Réaumur, a best-seller.[45] In the judgment of Paullian, La Pluche metamorphosed nature into affecting sights. The gendered language of feminized display was further manifest in the simile of an open book "exposed for all eyes to see."[46]

Following Fontenelle's lead, La Pluche employed the *galant* formula of intimate presentation. Readers were conducted into the firmament and the bowels of the earth through the lively conversation of four characters on an excursion. The young Chevalier de Breüil, Monsieur le Comte de Jonval, the Prieur de Jonval, and, somewhat tardily, Madame la Comtesse de Jonval, engagingly discussed how natural phenomena offered pleasure without the pain of pedantry. Influenced by the affective power of English descriptive poetry and believing that reason must be sensualized, La

Pluche claimed no one was immune to the feelings aroused by God's wonders.[47] Harnessing such noble tendencies, however, required the epitomization of the learned memoirs of Malpighi, Redi, Willoughby, Leeuwenhoek, Derham, and Grew in order to make them palatable. The boredom and banality of reality became dramatized through the introduction of emotion. Technical writing was converted into a cinematography of affective scenes. La Pluche should be credited with the invention of the art and entertainment format. In his distinctively modern brand of information technology, science popularization emerged as sugarcoated spectacle.

Instead of passing methodically from grand universals to insignificant particulars, and so obeying the inexorable logic governing a scientific report, La Pluche began with objects close at hand, small enough for a child to hold. Tiny animals, plants, insects, shells—all capable of being easily collected during a stroll—composed that gentler path to knowledge he termed "la physique des enfans." The spectacle format was inseparable from innocent and feminized diversions that strayed from the straight and narrow route of serious adult male scholarship. The rhetoric of display presented "solely the exterior or that which strikes the senses." Like Nehemiah Grew's concrete rendition of whelks in terms of the sea snails' striated and stippled carapace (fig. 160), La Pluche stressed "the outside of nature." Declaring that "the spectacle is for us" and that it consisted in "the superficial decoration of this world and the effect of its machines," he left the "delving task" to "a superior order of genius."[48] Unlike the hard words of later quantifying observers or the vulgar shoptalk of drudging tradesmen, the aristocratic dialogue engaged in during a leisurely walk was patently unproductive.[49] Looking at the glistening head of a fly or examining the needlelike sting of a bee under a microscope led to a purely aesthetic judgment. The limits of human skill and the crudeness of man-made instruments were starkly contrasted to the visible splendors of the universe created by the divine artisan.[50]

The great entomologist and technologist René-Antoine Ferchault de Réaumur admired La Pluche's extracts and envied his novelistic style. Nevertheless he wanted the six-volume *Mémoires des insectes* (1734–1743) to go beyond exhibitionism and be useful to French industry. Curiosity about the spectacle was to be harnessed to the enterprise of making discoveries. Drawing on a practical psychology of perception also motivating inventors like Vaucanson and entrepreneurs like Duchesne, Réau-

Tab. 10.

Square Wilk 1     2

Long Square Wilk 1     2

Thick Lipp'd Wilk 1     2

Triangular Wilk 1     2

Inverted Wilk Snail 1

160

Nehemiah Grew

*Whelks*

1681

from *Musaeum Regalis Societatis*

161

René-Antoine Ferchault de Réaumur

*Moths and Holes Chewed in Cloth*

1734–1755

from *Mémoires pour servir*

*à l'histoire des insectes*

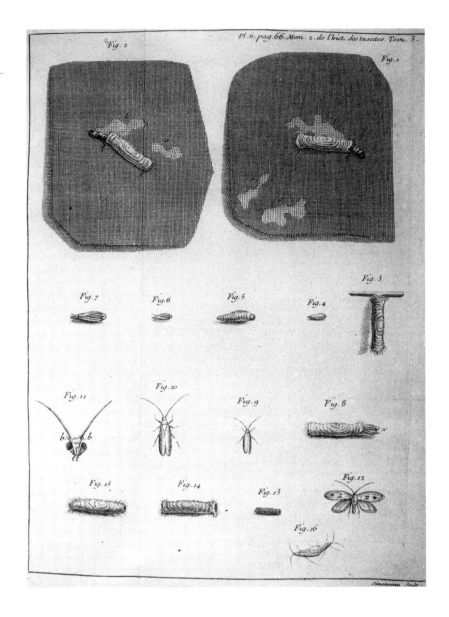

mur (1683–1757) remarked that utilitarian processes often originated in "what pleases and amuses." His lifelong love affair with lowly insects and their commercializable products led to the composition of an illustrated tribute he hoped "the entire world" might wish to consult. "The taste for the marvelous is a general taste, it is this taste that drives the reading of novels, short stories [*historiettes*], Arabian tales, Persian tales, and even fairy tales, rather than true histories [*des histoires vrayes*]." Yet Réaumur insisted that the realistic account of insect behavior contained more "de merveilleux vrai" than any other kind of biography.[51] He mourned the fact that few such studies existed in French or even other vernaculars. Those that did—such as the catalogues of John Ray—were unfortunately composed in a tedious style pleasing only to those who already adored the subject.

Réaumur's exhortative and sensuous brand of natural history fell in between La Pluche's rococo optical eroticism and neoclassical nongraphic descriptions or abstract taxonomies. A *méchanicien* of the highest order, this mathematician, natural historian, experimental physicist, and consultant to France's iron manufacturers and silk factories was also a member of the Royal Society, the Prussian and Swedish Science Academies, and the Institute of Bologna. His investigations into the morphology of the humble caterpillar conveyed excitement for its aesthetic or formal properties as well as for its contribution to the international trade in necessities and luxuries. Citing the honey and wax generated by "industrious" bees and employed to fabricate lacquer, varnish, and Morocco leather, or the cochineal scales crushed to procure red dye, he believed there were many analogous discoveries just waiting to be made. From the dream of flies cross-pollinating domestic and wild fig trees to the frustrating search for insecticides, Réaumur alternately seduced and horrified his well-to-do readers. Drawing harrowing pictures of destruction to galvanize future landowners into action, his illustrator, Mlle Dumoustier de Marsilly, and his engraver, Philippe Simonneau,[52] produced inflammatory images of minute marauders gnawing holes into plants, trees, fruits, grain, furniture, ship planks, clothing, and fur (fig. 161). Even human bodies were not immune to infestation by lice, fleas, and mites.[53]

The *mémoire,* then, shared with the *spectacle* the theatrical tactic of highlighting singular events as discrete displays. Like Lister's conspicuous framing of diluvial remains, moths were rendered extraordinary by an isolating *montage* that magnified their exploits. Such partial or cell-like scenes differed from La Pluche's seamless succession and distant perspective, however, by inviting viewers to engage in close-up observation. Hairy chrysalises and bristling cocoons reminded young and old alike of their dual role as admirers and consumers, onlookers and producers of perishable goods.[54] Réaumur offered an alternative to the unrestricted gluttony of dazzling spectacle and to the modern technocratic devouring "gaze from nowhere."[55] His undisguised enthusiasm for insects allowed him to see from their perspective, to understand their instincts, while situating that knowledge within a human context requiring the transformation of contemporary practices. Vision was cognitive, just as images were about demonstration, not satiation.

### The Polymathic Cabinet of Curiosities

While much has been written on the wonders housed in seventeenth-century museums,[56] I want to move the interpretation of entertaining oddities in a new direction. The analogy of the world to a picture also governed the treasury of curiosities. Splashy arrangements turned the repository into a *theatrum* whose grotesque materials were accumulated in order to be looked at.[57] Amusing medleys comprised *pittoresque* compositions on a grand scale. Bewitching arrangements of colorful rough stuff, like *pittoresco* unfinished brush strokes,[58] piqued the curiosity of the public. As a mismatched compositional impasto, striking lapidary fragments stimulated the viewer to piece together the puzzle. The dexterous, indeed acrobatic distribution of stony irregularities into apparently haphazard juxtapositions not only enticed the eye but stimulated conversation. Crumbling shells, clumps of madrepores, coral branches, miniature busts, Chinese porcelain teapots, small medals, intaglio gems, pottery shards, drawn and engraved portraits, masks, carved ivory, pickled monsters, religious utensils, and multicultural remains cacophonously "chatted" among themselves and with the spectator (figs. 155, 162). Like shapeless pigment stains or confusing blots, their manifest incompleteness precluded incorporation into a seamless narrative and controlling taxonomy. Delighting the amateur while defying the classifier, these collections were anamorphic.[59] Perplexing contents awaited resolution in the delectating vision of the beholder.

Crammed shelves and drawers, with their capricious jumps in logic and disconcerting omissions, resembled the apparent disorganization of talk. Exhibitionist dialogue taunted and parried. Like the gymnast's ostenta-

162

Salomon Kleiner

*Front Section of Open Display Cabinet*

1751

from *Christophori de Pauli Pharmacopoei Camera Materialium*

tious routines and the magician's seductive flourishes, snatches of information were volleyed in a spatialized give and take. Words and images flew without revision and erasure. Aposiopesis, recently found to characterize Sterne's *A Sentimental Journey* (1768) and Goethe's *Die Leiden des jungen Werther* (1774), was precisely such a rhetorical strategy for mimicking the fluidity and randomness of speech patterns. Dashes, ellipses, half-finished sentences, and even silences symbolized freedom from the literary constraints of syntax and punctuation ruling the eighteenth-century novel.[60] I suggest that this fragmenting oral-visual composition of a text—done ironically to portray the inadequacies of writing—was precisely the same broken pattern governing the unblended items in the polymathic *Wunderkammer.* Forms without geometrical form were "illiterate." Ruins were runes, resembling those barbarous and forgotten letters decorating eroded ancient monuments. Effaced by time and obliterated from human memory, worn-out inscriptions became empty calligraphy and meaningless ornament. Endlessly fascinating to eighteenth-century virtuosi, spotty debris spoke not with the clarity of a classical language but in "a more obscure dialect."[61]

Because sketchy oddities had lost, or never possessed, a legible symmetry and formal regularity, they lacked a preestablished symbolism. Disorderly and apparently useless, they encouraged reading in. The viewer was compelled to fill in the gaps through imaginative projection. Recall how the representation of unspoken speech taboos similarly drew viewers of rococo paintings into the game of filling in the blank. Completing the gesture was like providing the missing word in a conversation. The viewer/speaker assumed a responsibility in the creation of the unstatable witticism (see figs. 146, 147, 148). The illusion of intimacy manufactured by Greuze or La Tour was closely allied to the spectacle mode. Like La Pluche's familiar conversations, conveying the feeling of direct access to nature, uncensored communication between complicit partners produced a sense of unfettered immediacy.

The desire magically to complete the incomplete belonged to the same baroque mentality that conceived of painting as theatrical and theater as pictorial.[62] Both genres pleased the senses by making ideas concrete in a kinetically conceived medium. Salomon Kleiner's (1703–1761) album of watercolor interiors of the "Red Crayfish" pharmacy and private museum in Vienna wonderfully instantiated the belief that to think is to paint (fig. 163). This important topographer and portraitist of royal and

CHRISTOPHORI DE PAULI
PHARMACOPOEI
CAMERA MATERIALIUM
AD VIVUM DELINEATA.

163

Salomon Kleiner

title page to *Christophori de Pauli Pharmacopoei Camera Materialium*

1751

princely palaces[63] captured the stagelike presence of the monumental cabinet of curiosities located at the heart of the dispensary (see figs. 155, 162). With its doors flung open, the dramatically enframed *Wunderschrank* offered a spectacle on all four sides. The illusion of a wraparound proscenium demonstrated the tenacity of the pictorial paradigm in Catholic countries before the expulsion of the Jesuits.

Christoph Laurence Joseph de Pauli's elegantly spacious and frescoed druggist's shop was captured in Kleiner's graphic series as a complex environment where economic and aesthetic practices commingled (fig. 164). Useful medicine and physic rubbed shoulders with promiscuously mixed anomalies, gems, and orfevrerie. The ordinary commercial transactions between transitory customers, owners, and hired help intersected with the extraordinary microstructure of a permanent collection of rarities. If the bustling business world reified social hierarchies, the enchanting *Wunderschrank* elided differences among media, combined the natural with the artificial, and juxtaposed the mechanical with the organically grown in the manner of a present-day gridded digital image. The exotic antecedent of miniaturizing "electrobricolage," fabricated today from quotidian files, desk litter, and the detritus of cyberspace,[64] Kleiner's unique suite of interior views provides a glimpse into how the mid-eighteenth-century middle- and upper-class consumer might recombine the auratic objects amassed in a *camera materialium.*

The unpublished and unengraved drawings are of special interest because they document the complex layout and multiple uses of this aromatic dispensary. Part Baconian laboratory/workshop/sales premises, part library/"chamber of artifice"/collection of specimens (fig. 165), it contained also a window alcove (if not a hall) of models, machines, and inventions (fig. 166).[65] The de Pauli family acquired the pharmacy in 1712. Presumably they began gathering artifacts then, but this is uncertain since items may also have been inherited from an earlier generation living at the time of the Catholic campaign against the Grande Porte marked by the victory of John III Sobieski.[66] A late flowering of the mannerist *Wunderkammer,* this private gallery contained a typically eclectic mixture of the fine and decorative arts, antiquities and petrifactions, Turkish daggers and *spolia* dating to the lifting of the siege of Vienna in 1683, crucifixes (fig. 167), votive hands, portrait miniatures, caricatures, exotica, erotica, *memento mori,* narwhal tusks, ammonites, and armadillo carapaces. If this principle of restless irreducibility emerged from a Renaissance and baroque poly-

164

Salomon Kleiner

*Interior of Pharmacy*

1751

from *Christophori de Pauli Pharmacopoei Camera Materialium*

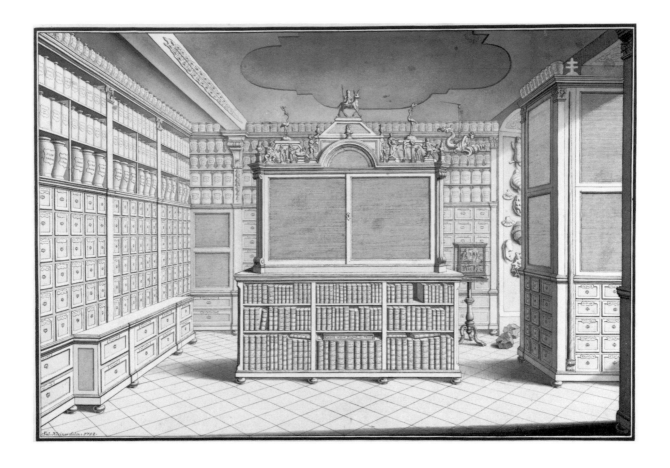

165

Salomon Kleiner

*View into Museum with Display Cabinet Closed*

1751

from *Christophori de Pauli Pharmacopoei Camera Materialium*

166

Salomon Kleiner

*View of Window Bays (with Air-Evacuating Pump)*

1751

from *Christophori de Pauli Pharmacopoei Camera Materialium*

167

Salomon Kleiner

*Side Elevations of Display Cabinet (with Crucifixes and Automata)*

1751

from *Christophori de Pauli Pharmacopoei Camera Materialium*

mathy, then the attention paid to the business of daily life belonged to the eighteenth-century empiricist rhetoric of demonstration.

The magical display cabinet dominating the sanctuary of the back room was organized according to the principle that, while material phenomena might be made to correspond, they could not be translated into one another. Small, tactile objects existed within a sensual sphere of analogies in which they served both as tokens for, and as reproductions of, an otherwise unseizable realm. The homeopathic dispensary in the front room flanked the street and was presided over by a portrait of Theophrastus Paracelsus. Stocked with savory condiments, pungent herbs, and aromatic botanicals, it was the zone for *practical* curiosities (see fig. 164). Whether under the eye or under the nose, shown in a treasury or stocking a utilitarian space, miniature goods actualized the hermetic connections existing between macro- and microcosm, body and mind.

The polite conversation of deferential purveyors with customers of good breeding was mimicked at the inanimate level in the mannerly interactions among disparate items in the display cabinet (see figs. 155, 162). In both the dim pharmacy and the well-illuminated *camera materialium,* social symmetries and decorum were maintained among people of distinction and objects of quality. In these potentially disruptive settings, individuals were permitted to be proximate but not to jostle one another in an uncivil fashion. The simultaneously magical and workaday world of the druggist's shop, paneled from stone floor to airy vault with drawers and jars, was otherwise sparsely furnished. An elegant porcelain stove faced an elaborate openwork rack suspended from the ceiling. Festooned with balances, this dangled over the heads of apothecaries filling prescriptions behind a long, L-shaped wooden counter terminating in columns bearing mortars and pestles. In the sales area, as well as in an adjacent room, apprentices pulverized pills, weighed powders beneath a painting of Christ the Healer, and shaped artful sugar cones. These tradesmen and their clients existed within the same vitalistic universe that the collection of rarities metaphorically evoked.[67] Predetermined social structures and prearranged artifactual patterns were enlived by a network of forces and connective energies.

Such dynamism and diversity demonstrated just how thin the line was separating orderly from disorderly behavior at both the animate and inanimate levels. Kleiner's topographical record of de Pauli's frescoed modern dispensary and heteroclite cabinet of ancient remains captured the polymathically conceived universe as it was about to be eclipsed by static and nonvisual methods of classification. After 1750, such pleasing spectacles of *Gesamtkunstwerk* and syncretic taste were judged useless, fatiguing, and ostentatious.[68] Their disappearance signaled the demise of the baroque ideal of a combinatorial art and, with it, of the showy *polyhistor*. The transformation of aestheticism into criticism and archivism was prompted by the realization that material multiplicity could no longer be circumscribed in a single system of etiquette.

*Pansophia,* or the amassing of things based on their interesting looks, was reified specifically in de Pauli's corner arrangement of unidentifiable objects waiting to be intellectually discovered (fig. 168). These shapeless remnants, lining the shuttered wall units set into the window bays of the museum (see fig. 166), presented a visual riddle. The amorphousness of brittle fragments was accentuated by the regimented procession of apothecary jars that loomed crushingly over fragile detritus. Old and new mentalities collided in a single space. Identical rows of large, shapely amphoras defied that fundamental premise of *Wunderkammer* decorum whereby even the most minute and irregular specimens or artifacts became perceptually piquant through picturesque distribution. Five rows of stately vessels methodically vanquished an eight-tiered mosaic composed of tiny and confusing details still awaiting an epistemological justification. This purified geometry was only apparently softened in the varied ranks of labeled square, rectangular, oblong, and cylindrical drawers and bottles ornamenting the exhibition area (see fig. 165).

If the classicizing march of standardized containers around the room resembled the even distribution of letters across a printed page, the medley of farfetched substances imitated conversation. Meaning accrued around heterogeneous items only in relation to a discussion that moved from diversion to knowledge. This rococo fluidity was fundamentally at odds with the abstract geometry of a limited number of legible forms central to *fin de siècle* neoclassical architectural and decorative schemes (fig. 169). In this artificial language, the character of even the smallest architectonic detail was intelligible because it was grounded in Platonic Ideas that were above, and prior to, any material medium.[69] Like unrigorous philosophical eclecticism and the sophistic orator's *copia rerum,* polymathic *mélanges* came to be disciplined by a reductive systematics. On one hand, then, de Pauli's theater of curiosities reflected an old-fashioned

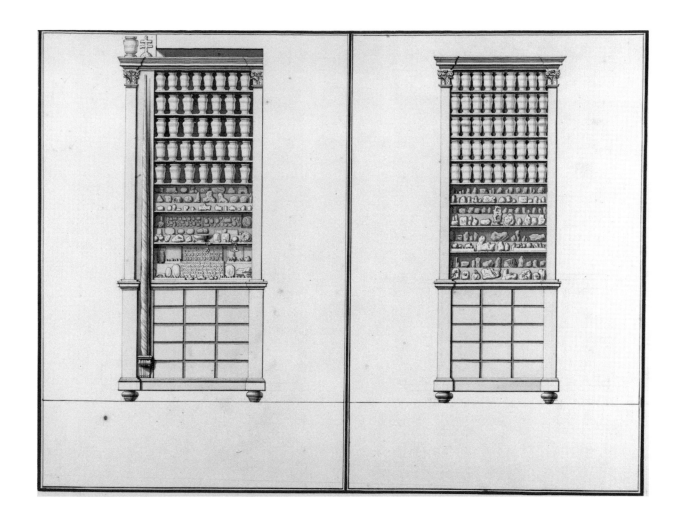

168

Salomon Kleiner

*Shapely and Shapeless Objects*

1751

from *Christophori de Pauli Pharmacopoei Camera Materialium*

169

Salomon Kleiner

*Ranks of Apothecary Jars*

1751

from *Christophori de Pauli Pharmacopoei Camera Materialium*

Leibnizian optimism. The ingenious collector, as well as the apprehending minds of equally individual beholders, coordinated a perplexing plurality of monads.[70] Harmony emerged when an ephemeral aggregate could be combined together into a mental image by the viewer.[71] On the other hand, this sanguine outlook, predicated on the compossibility of simple substances, was increasingly challenged by a skeptical pessimism that no longer believed appearances or disciplines might be optically conjoined.

In sum: bereft of labels and shapeless by any classical canon, de Pauli's lithic marginalia were doubly ruinous. Unmoored from a past context, this sea of fragments was incomprehensible and, so, grotesquely *modern*. That is, by not possessing a preestablished meaning, these shards were new because open to continual interpretation. Existing both as artful sketches, perpetually in the state of *non finito,* and as allusive natural remnants, these unreadable details belonged to a totality forever evading the spectator. Rubbish to systematic collectors of antiquities and natural history specimens, ambiguous curiosities resembled rumors. Being without discernible inscription, such barbaric bits and pieces were illiterate and could only whisper indistinct or garbled messages down the centuries. Lacking the clarity and distinctness of public pronouncements backed up by a recognizable *auctoritas,* broken stones were like vague opinions or snatches of muttered speech.[72]

Nature's *joli* ornaments thus resembled the intimate organic novelties manufactured by rococo artist-artisans. This decorative and useless pot-pourri of concrete media, addressed to wealthy consumers, was part of a sensualized interior ensemble whose diminutive and "feminine" paintings, statuettes, porcelains, mirrors, jewelry, and shells also came under attack in a sterner age.[73] The demise of cabinets of curiosities coincided with the fall of the artisan. The demotion of trade painting in the second half of the eighteenth century, or the daubing of signboards, coaches, houses, and stage scenery, was the result of the academic insistence on the conceptual nature of art.[74] Just as the guild mentality fostered manual skills and valued the material pigments and stains composing a physical medium, the *Wunderkammer* mentality favored colorful agglomeration rather than rational design. Singular works of artifice and nature were part of a sensory realm grounded in appetite, not in superior knowledge based on ideas invisible to the untrained beholder.

Long before the surrealists, convulsive beauty resided in what Sigaud de La Fond termed "extraordinary eccentricities."[75] The French popular science demonstrator's *grotesqueries* were no longer housed, however, in a cabinet where they might imitate in miniature an earth bustling with myriad forms of life. Instead, *écarts* were abstracted from the living spectacle of nature and gathered into a dictionary. Inquisitive readers, consulting the *Merveilles de la nature* (1781), were enticed into searching for ways of turning mysterious debris into utilitarian objects rather than developing into curio-hording plutocrats.[76]

Exactly one century earlier, the Presbyterian physician and botanist Nehemiah Grew (1641–1712) found polymathic collections to be short on serious and intelligent reflection about what museums were actually for. Like "popish" miracles, these encyclopedic exhibits of curiosities inspired wonder and empty gaping. Perceived now as the result of thoughtless rummaging, such surfeit became demoted from tasteful display to lowbrow mass medium. In his catalogue of the museum of the Royal Society, housed at Gresham College in London, Grew expressed what would become a growing disdain for flashy exotica. Following the tenets of Baconianism, he believed that not only the contradictory, equivocal, and strange constituted the physical world but that the ordinary had a presence too.[77] Protestant mercantile morality infused his call for exact inventories of common goods as opposed to the wasteful accumulation of preposterous *bizarreries*. The *Musaeum Regalis Societatis* (1681) indicated how the display of preternatural rarities was to give way before the anatomization of neutral matters of fact. The superficial exhibitionism of acquisitive virtuosi, allied to equally suspect chiromancy, geomancy, magic, and hermeticism,[78] was to be unseated by deeper methods of ordering.

As Secretary of the Royal Society since 1677, Grew spoke of linguistic categories and textual descriptions as means for getting at the more important and noncounterfeitable invisible reality underlying natural products. "Names of Things should be always taken from something more observably declarative of their Form, or Nature." But these verbal epitomes must penetrate below the visible surface to comprise "a short Definition" summing up the essence of a physical object.[79] A catalogue was a nomenclature, systematizing the not always evident characteristics of plants and animals. "Diverse Wilks" (see fig. 160) offered a case in point.

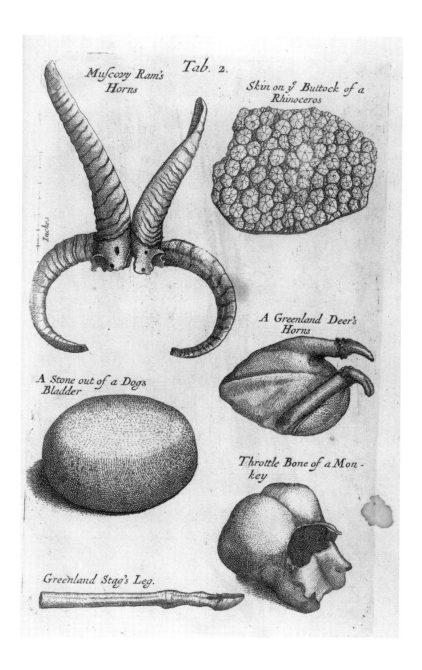

Part of the treasury of some 600 great and small shells in the Royal Society's collection, they were provisionally arranged according to the best method available at that moment. "Afterwards," he added, "[I] shall digest them into Schemes."[80]

To strip marvels of their former aura, the naturalist created visual "Lists" reducing potentially numinous wonders to materially interchangeable specimens for critical inquiry (fig. 170). Unlike de Pauli's polymathic miscellany, the magnified "Skin on ye Buttock of a Rhinocerous," the "Stone out of a Dog's Bladder," or a "Cat's-Tail Sponge" lost their optical allure and so ceased to invite the projections and stimulate the conversations of nonprofessionals.

The British botanist's demystification of cosmic analogies was prescient of the later eighteenth century's antiparticularist aesthetic. While Kleiner's album bore witness to the triumph of the undecidable and feminized detail,[81] constantly drawing attention to itself by its showy idiosyncrasy, Grew's downright homely and prosaic fragments were forerunners of a neoclassical methodism. Organic pieces were no longer pretexts for playful combinations but irreconcilably compartmentalized bits. Interpreting these atomistic distinctions became a male prerogative insofar as the importance of surface and depth, part and whole, matter and form, text and image had exchanged positions in the epistemological hierarchy.[82]

The growing tension between the conflicting claims made by the particular and the partial as opposed to the general and the total can be witnessed within a single famous institution. The struggle of Georges-Louis Leclerc, Comte de Buffon and Louis-Jean-Marie Daubenton to organize the "pittoresque fatras" of the *cabinet d'histoire naturelle* is a case study for the inadequacies of both de Pauli's juxtaposition of heaped-up inequivalencies and Grew's reductive language games, dissecting phenomena into finer and finer acts of differentiation.[83]

The Jardin du Roy, founded in 1635 by Guy de La Brosse, *médecin ordinaire* to Louis XIII, blossomed during the eighteenth century. In 1729, its medicinal and pharmaceutical mandate was expanded to embrace the three kingdoms of nature. This extension of mission beyond physic was reflected in the amplification of its collections and staff. Doctors, surgeons, and apothecaries were augmented by natural philosophers who did not belong to the medical world. Macquer, Fourcroy, Faujas de Saint-Fond, and Lamarck introduced chemistry and the earth sciences into the domain of horticulture.[84] In 1739, Buffon (1707–1788) became director and held the post for almost 50 years. During his activist reign, he also produced with Daubenton, Gueneau de Montbeillard, and the Abbé Bexou the monumental 15-volume *Histoire naturelle* (1749–1767), whose publication continued posthumously until 1804 through the agency of the Comte de Lacépède.[85]

Our concern, however, is with Daubenton's description of the *cabinet,* published in the first three volumes of the *Histoire naturelle.* In 1745, Buffon installed Daubenton (1716–1800) as *garde et démonstrateur du cabinet,* conferring on his compatriot the Herculean task of cleaning the Augean stables. Dominique Sornique (1708–1756) and Pierre-Edmé Babel's (1720–1775) headpieces for volume III, engraved after original drawings by Jacques de Sève,[86] give some sense of the transmogrification of "picturesque trash" into two orderly rooms. As the tiny vignettes show, authentic treasures still jostled pickled monsters (fig. 171), and modern technical marvels, like Vaucanson's *Duck,* continued to be enshrined alongside dubious antiquities (fig. 172).

The *cabinet* was open regularly to the public (symbolized by the gawking and conversing putti) on Tuesdays and Thursdays, except during vacation. Apparently it attracted large crowds. What the 1,200 to 1,500 *spectateurs* per week experienced, however, was the transitional moment when the old polymathic jumble was in the process of becoming a public service display requiring the interpretation of experts. Around 1750, Paris witnessed a decline in virtuoso or *curieux* courtier collections while interest grew among money handlers, the bourgeoisie, artists, and antiquarians. This shift, also mirrored in the move away from amassing medals to gathering natural history specimens, coincided with the formation of scholarly collections as opposed to those arranged solely according to aesthetic criteria.[87]

Daubenton's illuminating account of the problems surrounding the organization and exhibition of a heterogeneous natural history collection for a diverse audience was published in the *Encylopédie.* Like contemporary museum curators, he was forced to reflect how once-cherished personal possessions might become meaningful in a social environment. How can things become "good" goods, not commodities but intelligently internalized things shaped through different types of encounter by different kinds

De Save inv.                                                                 Babel Sculp.

171

Georges-Louis Leclerc, Comte de Buffon

*Cabinet du Roy*

1749

from *Histoire naturelle*

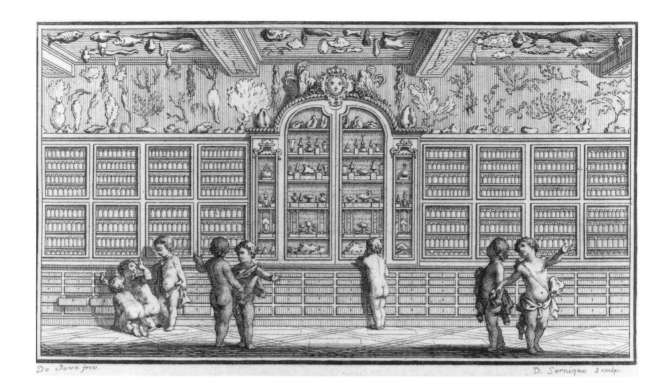

172

Georges-Louis Leclerc, Comte de Buffon

*Cabinet du Roy*

1749

from *Histoire naturelle*

of observers?[88] As chief curator, he was interested in discovering installation patterns that would stimulate viewer participation. He hoped to unfreeze both the beholder's preformed habits of looking and the object's intractability. The transaction was to be entertaining and educational since specimens could pleasingly modify the visitor's prior expectations and associations.

Daubenton distinguished the intimate *cabinet,* that private haven in a house destined for study or display of precious curiosities (see fig. 154), from the public *cabinet d'histoire naturelle.* Less grand than Pannini's basilican art gallery (see fig. 152), it was composed, nonetheless, of a suite of rooms: of which the largest had to contain a variety of collections exemplifying the multiple productions of nature. Reverting to the pictorial metaphor of the frame—central to the enterprises of Lister, La Pluche, and Kleiner—he asked rhetorically what means might be employed to convert a chaotic assemblage into a comprehensible spectacle. Gathered together into a varied and nuanced tableau, animals, plants, and minerals located in the same space and "seen, so to speak, at a glance" needed to convey a proper idea of themselves to the viewer. This painterly method of display resembled Lister's illusionistic strategy of turning even the most modest shell groupings into a pretty picture (fig. 173). Such pictorial glamorization of specimens was patently unlike Grew's borderless, and hence "naturalistic," plain marine vegetation that coexisted unassumingly on the same neutral page (fig. 174). Reminiscent of La Pluche's epitomization of earth and sky, Daubenton compressed three-dimensional objects into a sketchy *abrége* (see fig. 171).[89]

The compositional conundrum of integrating the many into the one was especially acute in the capricious category of the *règne animal.* Its swollen ranks included bones remarkable for their fractures, deformities, and rickety pathology, wet and dry anatomical preparations, fetuses at different stages of development and other "singular" specimens preserved in alcohol (fig. 175), beautiful wax and wood models, mummy parts, stones removed from human organs, skeletons of quadrupeds, innumerable shells, insects, butterflies, and sea or land plants. Preserving these irregular and corruptible substances drove Daubenton to distraction. He urged instrument- and glassmakers to manufacture vessels that were longer and taller so that they might comfortably house small quadrupeds (fig. 175, no. 7). A need also existed, however, for slender, cylindrical tubes to encase fish, lizards, and snakes without leaking. Alluding to the difficulty

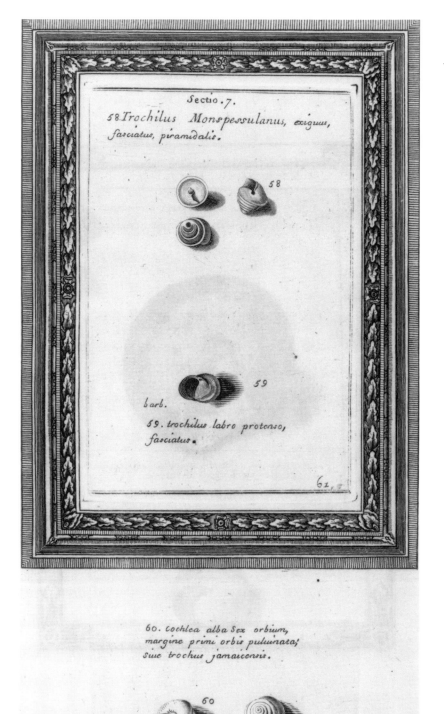

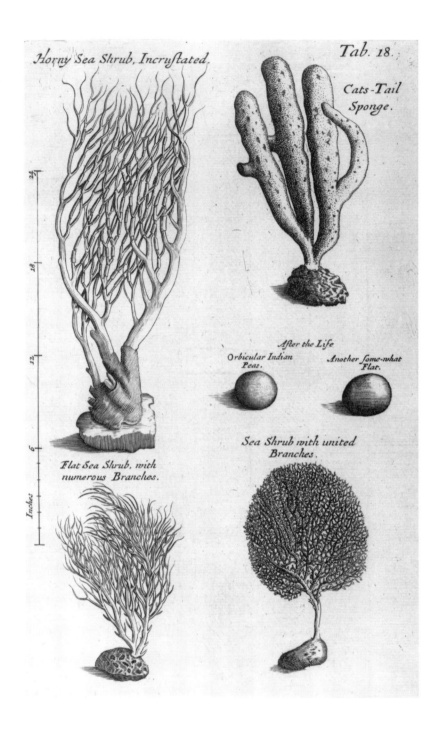

Horny Sea Shrub, Incrustated.

Tab. 18.

Cats-Tail Sponge.

After the Life
Orbicular Indian Peas.
Another some-what Flat.

Sea Shrub with united Branches.

Flat Sea Shrub, with numerous Branches.

175

Georges-Louis Leclerc, Comte de Buffon

*Pickled Monsters*

1749

from *Histoire naturelle*

of tightly sealing oddly shaped jars, especially when they had to be laid on their sides in crowded shelves, he called for the invention of an impermeable mastic composed of mercury and lead to prevent spillage, evaporation, and eventual decay.[90]

As an economy-conscious realist, this custodian of perishable goods remarked that in an established museum one had to use and reuse any container one could lay hands on since the universe was filled with strangely shaped organisms that required bottling. Conserving specimens, especially during the "five months of torment" stretching from April to September, meant the curator did the best he could under adverse circumstances. Like Réaumur, Daubenton conjured up the specter of countless and prodigiously multiplying invisible insects destroying dessicated preparations before their vigilant guardians even knew they existed. Describing the museum director's nightmare, Daubenton itemized a terrifying troop of "worms, beetles, moths, butterflies, mites, etc. . . . They gnaw the flesh, cartilage, skin, fur, & feathers; they attack plants, although dried with the greatest care; one is aware that wood itself can be reduced to powder by vermin."[91] This thoughtful natural historian, then, also sheds light on the far from ideal conditions determining eighteenth-century public displays. He painted a warlike picture of beleaguered staff battling the recurrent problems of nonstandardized organic forms, inadequate facilities for preservation and climatization, and a perpetual lack of sufficient exhibit areas. Inadequate space meant that items often had to be stationed vertically or horizontally, not because of theoretical dictates but because of the limitations of the site.

This same no-nonsense attitude governed his revealing discussion of which installations were best to encourage the study of natural history. Continuing his artistic analogies, Daubenton argued that each item should be "framed" under glass and given its own label. He realized, however, that individual objects were part of the total *cabinet.* This meant they could not be heaped up without method and taste in the manner of phenomena randomly encountered outside its walls. The perspicuous curator, like the knowledgeable collector, needed to know what to keep or discard and how to impart "a fitting arrangement" to each item. Paradoxically, the system of classification shaping a *cabinet d'histoire naturelle* was not natural. The universe possessed "a sublime disorder," while the purpose of the museum was to instruct the public in an orderly manner. In the artificial setting of the natural history collection, the viewer must be able to discern, in detail and in a logical way, what the world presents *en bloc*. Artful exhibitions, therefore, depended on displaying nature's treasures according to some subjectively determined pattern. Such artificial taxonomies could be based on the greater or lesser importance of the species, the viewer's interest in them, or some other scholarly reason.

Voicing the dilemma faced also by late twentieth-century museum directors, Daubenton lamented that it was difficult to find a single organizing principle that pleased "gens de goût," interested the "curieux," and provided insights to "savans." Thus the tendency was to emphasize either the aesthetic or the scientific goal. The absence of systematization in many private European collections, he believed, resulted in the drawback that a multitude of possessions were chaotically "thrown as into a well." At the same time he pointed to the heterogeneous public thronging into the *cabinet du roi*. Coming from all walks of life and from countless foreign countries, these *spectateurs* arrived in such droves that, in fine weather, there was scarcely enough room to accommodate them. The problem was that since anyone could enter desiring to be amused or instructed, public institutions had a special obligation that owners of personal galleries did not. Seeking balance, he wanted to avoid both the pedantry shocking *honnêtes gens* and the fairground charlatanism impeding scientific research.[92] To attain a truly general appeal, the display must assist each visitor to glean information "at a glance," not only about the object but about its relation to others in the exhibition. Passing from one realm to another "without fatigue or the disgust arising from confusion," the beholder "engraved" the image of nature as a whole on his or her memory.

Ever the pragmatist, Daubenton lamented that extremely large items, such as mammoth bones and elephant tusks (see fig. 171), necessarily interrupted the overall classificatory pattern. It was difficult to find a spot for colossi as well as for minutiae, especially since objects from different realms were, in principle, not to be confounded. Despairing of perfection, no doubt precisely because of such practical predicaments, he claimed that the organization that satisfied the mind almost never pleased the eye. Dangerously splitting apart cognition from perception in the attempt to provide a counterbalance to the new rage for system, Daubenton suggested there might be purely optical arrangements. The surplus or nonessential portions of the collection could be distributed solely according to the aesthetic dictates of symmetry, chiaroscuro, and contrast (see fig. 172). Indeed, he claimed partially visible objects were always the most "piquant." Specimens that were "skied" or randomly strewn about in favorable viewing locations stimulated visitor interest by attracting attention to the "spectacle." He felt that such an ensemble "agreeable to the eyes" pleased the majority of beholders because its contents were not constrained to obey a rigorous method.

While Daubenton was troubled by disorder, he feared the consequences of an inflexible systematization more. He worried that museumgoers who continually observed *naturalia* predetermined by someone else's theory of classification might themselves neglect to study nature. Arguing for a "natural" arrangement, that is, one most akin to phenomena *en liberté* before they were confined in a small space, he wished to mitigate the power of artificial taxonomies. There was room for the aesthetic because even the most capacious treasury could not display in miniature all organic progressions, and ceaseless acquisition entailed the constant displacement of existing specimens. Declaring that something valuable was always learned during the manual and visual activity of installation and reinstallation, Daubenton claimed this dictum held true for those ocular combinations and recombinations made "only for pleasure."[93]

This unresolved dichotomy between a rationally ordered repository of knowledge and a free-form entertaining spectacle was recorded in de Sève's headpiece (see fig. 172). Standardized faience pots containing immutable minerals vied with brittle corals and rotting reptiles and amphibians defying any physical or intellectual frame.[94] The problem of what a collection should be at midcentury resulted from competing concerns and audiences. How, and for whom, could one achieve a totalizing ensemble without suppressing the individual parts and vice versa? The romantic generation continued to be vexed by the quandary of reason's demand for unity in the face of nature's manifold multiplicity.[95] Today we still live with the dualism pitting an "aesthetic" display, whereby purposeless objects are left to speak for themselves, against the demands of an overwhelmingly *textual* documentation. Contextualization too often means that artifacts of little intrinsic merit are put in the service of a theoretical distribution as tokens of an immaterial age, culture, or social system.[96]

## Album of Minutiae

Daubenton's compartmentalization of exhibitions into instructive and entertaining areas was swiftly and violently challenged by the French Oratorian Joseph-Adrien Le Large de Lignac (1710–1762). His *Lettres à un amériquain sur l'histoire naturelle de Mr. de Buffon* (1751) manifested the zeal with which the systematic model was embraced after 1750.[97] This distinguished naturalist, whose research on water spiders was sufficiently admired by Réaumur to be inserted into the *Histoire des insectes,* had only scorn for Daubenton's suggestion that certain genera eluded classification, that some species thwarted methodical distribution, and that others could only be included "pell-mell" because of their beauty, magnificence (*éclat*), or costliness. Lignac mockingly treated these charming, convention-defying exceptions as expensive pieces of furniture decorating the homes of wealthy amateurs who valued "an elegant disorder." He asserted that, on the contrary, there was nothing confused or equivocal about nature's series. This Cartesian maintained there was no Lockean beholder's share. The great chain of being was fixed and the differences characterizing each gradation did not depend upon our perception of them. Therefore even the "worst method," be it the alphabetization of the dictionary, was preferable to the loveliest symmetrical arrangement.[98]

With Lignac we can pinpoint not only the emergence of an antivisual rhetoric arguing that elucidation occurs in words not images, but also the neoclassical concern with a sequential narrative composed of isolatable facts. This "philosophic" naturalist had no patience with Daubenton's "aesthetic" concerns, still part of the lingering seventeenth-century heritage bequeathed by the virtuoso and polymath.[99] Instead of tasteful spectatorship, he sought rational structure rooted in rigorous observation. Order signified the precise separation of animals, plants, and minerals. It was no longer the amateur's loving and unifying glance that prevailed in the dissecting comparative method, but the professional's analytical *coup d'oeil.* Lignac also dismissed the concrete problem of fluctuations in scale that had vexed Daubenton. Relying on the example of books, he noted that libraries solved the quandary of where to store folio editions by an arbitrary fiat. Like the contemporaneous treatises on the art of experimentation discussed in chapter 3, Lignac believed that, occasionally, the classifier was entitled to violently interrupt an organic series. This artificial wrenching was justified by the conviction that phenomena were inherently mute and nature had to be forced into giving an account of herself.

Deriding Daubenton's rococo sensibility, reflected in the call for clean, well-lit, and elegant installations furnished with capacious cupboards, ample desks, and sparkling mirrors, he charged that the desire for a pleasing *ensemble* thwarted serious scientific inquiry into minute particulars. Pandering to the multitude, Daubenton's chaotic layout substituted entertainment for instruction. The custodian of the *cabinet d'histoire naturelle* was no better than the mannered landscape gardener who arranged exotic flowers into an enameled parterre to seduce the viewer by a riot of colors without any regard for lawlike principles.[100]

The shift from sensory impact to a rationalizing nomenclature was also a move from the extraordinary to the ordinary. Analysis meant that material things were decomposed into their normal or customary elements and then recomposed into a superior system knowable only through intellect, not perception. Daubenton as *garde et démonstrateur* at the Jardin du Roy inhabited the loosely constructed domain of the life sciences then in the process of becoming historical. Language theory, with its emphasis on itemization, classification, and logical narrative, increasingly pushed dilettantish visual discernment to the side. According to Duchesne's and Macquer's *Manuel de naturaliste* (1771), a printed explanatory guide was mandatory for the visitor to a picture gallery as well as to a cabinet of natural history.[101]

The new physiognomic criticism deciphered nature's charlatanic characters. Deep science probed beneath deceptive surface signs to excavate invisible typical properties. Rational archaeology stood in stark contrast to polymathy's shallow accumulations. The power of the mind supplanted the gullibility of the eye.[102] Similarly, collections became "organic," in the Kantian sense.[103] Individual components were ranked and graded according to an architectonic. Cabinets of curiosities, once socialized through spatial relationships prompting conversation, were demoted to being merely instrumental, functioning now as low-grade compendia of externally and mechanically related pieces. The expert, as Lignac declared, did not purchase the debris of great collections dispersed at auction. Unlike the wealthy amateur or commercial dealer, the specialist conducted research in the country where the specimens originated. "Not with money" but with observant "eyes that know how to see" could the investigator enter the international republic of erudite scientists.[104]

The English concept of field work, with its episodic gathering of bits of data, did not carry the French overtones of an a priori system capable of distinguishing between trivial and significant things.[105] John and Andrew van Rymsdyck's *Museum Britannicum* (1778) also formed part of the expanding community of knowledge about the earth. Unlike Daubenton's dichotomous exhibition space or Lignac's taxonomic suppression of merely visual features, these natural history draftsmen of Dutch extraction created an "ideal museum" by compressing fragments into an art book.[106] The holdings of the British Museum originated in the complex spectrum of diverse cabinets amassed by Sir Hans Sloane. Housed first in Great Russell Street and then moved to Cheyne Walk, this miscellany of antiquities, intaglio gems, medals, coins, and natural history specimens was arranged with some attempt at order. In his 1749 will, Sloane bequeathed the collections to the British people contingent upon the payment of £20,000 to his family.[107] Duly installed at Montagu House in Bloomsbury, the general public was educated and entertained by this unstable combination of *artificialia* and *naturalia*.

The Rymsdycks' paper "exhibition" of 30 plates not only broke the artificial sequence in which these antiquities and natural curiosities had been arranged when on display, but removed them from whatever order they had belonged to in the past. The selection, in the words of John van Rymsdyck, was made from "legacies bequeathed by rational beings to posterity [which] have been carefully preserved in the repositories of the British Museum." Given the vastness of that treasury, father and son decided neither to document its holdings nor to reproduce its "best" examples.[108] Instead, they transformed and compressed certain three-dimensional items into a two-dimensional epitome. The unique book of nature had literally become a reproducible digest containing abstracted works of art. "A variety of Picturesque, Curious, and Scarce Objects" were chosen with the intention of making them "instructive, entertaining, and useful." Like Daubenton, these artists were keenly aware that some pieces "will always be found more pleasing than others, according to the different tastes, studies, and geniuses of men." Bitterness accompanied this realization. Just as there was no single museumgoing spectator, there was no unified reading public. The awkward balance between entertainment and higher culture had been disturbed, as indicated by the change in terminology in the second edition replacing the more limited sense of *exhibition* with the notion of mass *display*.[109] Foreshadowing the nervousness evinced by late twentieth-century graphic artists in the face of electronic media

## 176

John and Andrew van Rymsdyck

*Grapholithi or Figured Slates,*

*and an Agat, with the Eclipse of the Sun*

1778

from *Museum Britannicum*

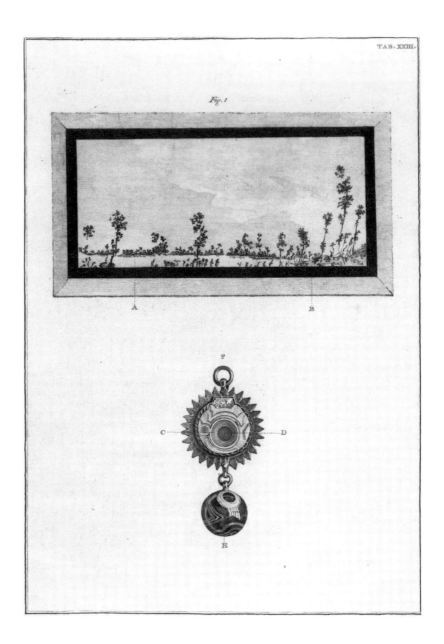

copying, they uneasily confronted a variety of specialized viewers, diverse technologies for reproduction, and competing interpretive communities.

When John van Rymsdyck showed their designs to potential subscribers, some "ladies and gentlemen" wished the illustrated book might be devoted to botany. Others expressed a desire for birds, butterflies, or quadrupeds. A few "wanted them all artificial." Deciding to commercialize the detail, he opted for "an Intermixture, which I suppose will consist of some things fine [fig. 176], others middling [fig. 177], a few *so so,* or perhaps but indifferent" (fig. 178).[110]

177

John and Andrew van Rymsdyck

*Hair Ball, Egg within an Egg, etc.*

1778

from *Museum Britannicum*

178

John and Andrew van Rymsdyck

*Coral "Hand," Encrusted Beaker*

1778

from *Museum Britannicum*

*Grapholithi, or Figured Slates,* an agate "with the Eclipse of the Sun," *Maccaw* and *Ivy-Owl's Eggs,* the hairball found in an ox's stomach, or even a *Coral Hand*—in spite of their superficial resemblance to Kleiner's record of mysterious and optically pleasing oddities—belonged to a different, isolating mentality. Declaring himself an enemy to "Nature Menders, Mannerists, or Antiques, etc.," this illustrator for the anatomical works of the celebrated Dr. John Hunter was similarly after the *true,* nonvisible character underlying different substances. His "new [artistic] Doctrine" was distinct from the "fashionable" improving and generalizing style of Reynolds currently in vogue. Apparently their designs had been criticized by "some" academic painters for being "too minute" and composed of "a great many little parts." The Rymsdycks' drawings, then, avoided the convention of rendering objects from a remote, spectacle-mongering vantage (fig. 179) as well as from an amateurish and approximating distance (fig. 180). Instead, they brought specimens "so near the Eye" that, rather than merely representing "the Effect of Nature," its particulars were seized in microscopic detail (fig. 181).[111]

Kleiner's illiterate debris (see fig. 168) and Lister's remarkable, but ambiguous, "sports" (fig. 182) were converted into abstract allegories wrested from their material contexts. No longer "talking" amongst themselves and with their audience, such illustrations already signified and were meant to be "read" by silent viewers. Cameolike, isolated images stood in naked relief against a blank page. Disproportionately magnified and hyperreal, the Rymsdycks' copies of figured stones and unusual items appeared simultaneously common and special, individual and universal. The illusionistic landscape automatically impressed by the "Hand of Nature" on a Florentine stone—with its low horizon, large sky, and picturesque clumping of foreground trees—out-Ruysdaeled Ruysdael. A fragment of nature simulating art imitating nature, this re-representation anticipated the refabrications of present-day exhibitions (see fig. 176). And the India agate, set into a pendant, looked forward to the era of museum replicas with its "true Representation of the Eclipse of the Sun and Moon on the bottom of which hangs an Onyx-Drop."[112]

Intensely focusing the viewer's gaze on a "Pregnant Egg," that is, one laid inside the other, or demonstrating how the strands of a hairball proceeded from deep inside the center of a hard and gluey substance, the artists made the ordinary appear extraordinary (see fig. 177). Conversely, an exotic coral was rendered familiar because "modled by Nature in the Form of a Hand

VAUXHALL GARDEN.

179

Augustus Charles Pugin and Thomas Rowlandson

*Vauxhall Garden*

1809

180

Maurice-Quentin de La Tour

*Portrait of Duval de L'Epinoy*

c. 1745

181

Louis Joblot

*Porte-Loupe and Specimen*

1718

from *Descriptions et usages*

*de plusieurs nouveaux microscopes*

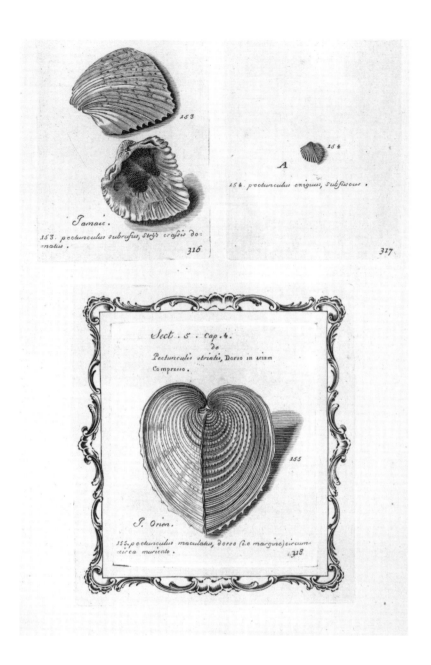

## 182

Martin Lister

*Pectunculi Striati et Maculati*

1770

from *Historiae Conchyliorum*

183

John and Andrew van Rymsdyck

*Encrusted Sword and Skull*

1778

from *Museum Britannicum*

or a Glove, with round Perforations," just as a tumbler encrusted with lime had its strangeness demystified by showing that "the Stone was once in a Liquid State, though some will have it to be made so by Fire" (see fig. 178).[113]

The two artists employed as engravers only "Men of Merit" able to execute prints that scrupulously copied the originals. Detesting the generalizing manner of using "Strokes, Hatches, and Gratework," the Rymsdycks de-

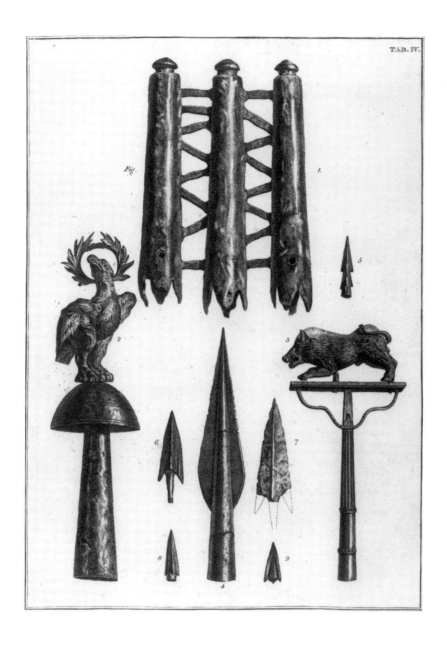

184

John and Andrew van Rymsdyck

*Roman Ensigns*

1778

from *Museum Britannicum*

manded a procedure as intricate as the objects represented. This God-like acuity in perceiving different textures (fig. 183) and variously wrought materials (fig. 184) was a corollary to the authors' belief that "Writing will never depict so strong an idea in the Mind, as a true Representation of an Object in a Drawing." If human vision is more or less blurred, depending on one's perspective, divine apprehension is perfect, complete, and always close-up.[114] Analogously, scientific precision in copying, unlike merely verisimilar illusion, made it seem as if artifacts demonstrated

185

Caspar David Friedrich

*Woman at the Window*

1822

themselves without the intrusive hand or eye of the historically situated observer. Yet simulacra, such as an encrusted iron *Sword and Skull* (with a portion of the humerus still grotesquely attached to the cranium), dredged from the bottom of the Tiber, or *Roman Ensigns,* unearthed in Scotland, served as reminders that perceptual distinctness could not guarantee cognitive clarity.[115] Indeed, the incongruity between prophetic vision and confused understanding became the romantic paradox (fig. 185).

This album of sharp but perplexing minutiae proved, if anything, that when we first encounter an event we may not understand it. Museum displays—whether de Pauli's paratactic juxtapositions, Daubenton's interrupted series, or the Rymsdycks' excerpted details—revealed that the beholder, like the traveler, began knowing something only the second time around.[116] To learn, viewers had to be enticed into re-presenting and reenacting in their own time and place what had been distantly accumulated.

Unlike today's false separation of works of art from technological inventions and popular imagery of all sorts, the eighteenth-century notion of an instructive, cross-disciplinary, and entertaining spectacle, based on a conversational give and take, needs to be brought back. This seems already to be happening in the chatty cyberspace of the "virtual gallery," where art in the form of digitized images in graphic files is shared through computer nets and webs.[117] But these disembodied on-line messages have no physical substance. Telepresence is no substitute for the wonder of material things experienced bodily and in common. This magic of living rarities must return to the desacralized and decontextualized space of the contemporary exhibition. Drama and ritual might undercut the bibelotization of art and nature as bric-a-brac.[118] Instead of the mania to pile up masterpieces like so many commodities, museums could help an increasingly fragmented and estranged society to visualize itself as a whole. In the brave, soon-to-be world of 500 cable channels, cultural institutions will have to demonstrate there is meaning beyond the media universe of advertisement. Learning from the Enlightenment, they could make us conscious that visual persuasion entails more than radiating bits of data.

## Looking Forward

To some, postmodernism has seemed the triumph of the phantasmic image.[1] To others, the late twentieth-century fascination with visual technology and multidimensional simulation is cause for suspicion and anger.[2] Jeremiads aimed at television's pollution of the symbolic environment and the nostalgia for a print-based epistemology[3] are disturbingly reminiscent of Enlightenment iconoclasm. From the theoretical perspective of Carrard, Senebier, Zimmermann, Lignac, or d'Alembert, not only was the baroque a civilization associated with absolute monarchy, luxury, mercantilism, and the alliance between church and state, but its political rituals and cultural metaphors were flamboyantly optical.[4] Amazing sights seemed dangerously akin to miracles, spells, and other fanatical delusions.[5] The disorderly evidence of the eye required chastening by language and conversion into rationalizable observations.

I have argued here and in *Body Criticism* that the Enlightenment was not just a figment of nineteenth-century historiography but existed as a well-defined period. Its coherence was stimulated, I suggest, by the threat of economic, social, and aesthetic instability posed by nontextualizable phenomena. Images, like conversations, proved maddeningly elusive and difficult to control. Power was at issue, especially when the producers of graphic designs claimed displays could be as instructive as texts and as entertaining as carnival shows. Nowhere can one see better the tension between information and misinformation than in optical "recreations," experimental trials, body tricks, and magical apparatus. In addition, illusionizing tendencies such as the spectacularization of nature and the pictorialization of museum installations were part of that larger technology of exhibitionism, and its countermorality, forming the theme of this book. These topics provide the central locus for examining a major issue they hold in common: the eighteenth-century obsession with false appearances. Was it progress or barbarism to arouse wonder by fooling people? What happens when utopian aspirations to perfect humanity through entertaining education confront a far from perfect reality?

In sum: self-consciously systematic, ethical, and linguistic, the Enlightenment intensified Descartes's conviction that error was the greatest evil.[6] Dedicated to the compulsive refashioning of the credulous and passion-ridden self, it was precisely that pan-European critical and pedagogical movement that defined itself by mounting a methodical attack against all forms of pseudos. Counterfeit presentations were synonymous with a corrupt "papist" oral-visual culture.[7] It has not been sufficiently emphasized, I believe, that the antiocular impulse of the early modern period grew out of a virulent anti-Catholicism. Analogously, for the neo-Luddites of the late twentieth century, a bewitching technology advancing mechanization, automation, and computerization troublingly defines our social environment.

While eighteenth-century readers certainly adumbrated the mass literacy drives of the nineteenth century, portraits and genre paintings by Maron, Lepicié, Boucher, and Chardin (see figs. 14, 17, 18, 40, 51) depicted the tactile joys of holding a book or perusing an intimate letter, not an imprisoning and voiceless typographical loneliness.[8] These representations of reading and writing as a novel spectacle demonstrate that early to mid-eighteenth-century literary consumers were still in transition from an oral to a silent mode of communication. They behaved as if they were beholders of living letters and listeners to incarnated words. As inhabitants of the *viva voce* realm of rhetoric, late baroque and rococo audiences had not yet severed deciphering from gazing. Contrariwise, living out loud was an impropriety among the splintered bourgeois publics springing up in the aftermath of the industrial revolution.[9] The "dire monotony of bookish idiom," in Thomas De Quincey's words, signaled the end of a rococo salon society with its *improviste* encounters.[10] It altered and radicalized the milieus of British coffeehouses and clubs. During the first half of the eighteenth century, before the proliferation of printed professional science manuals, these penny universities aspired to be neighborhood academies. Venues for lectures and theatrical courses given by "projectors," they drew all classes together through the staging of natural philosophy for group amusement and edification (see fig. 4).[11]

Henri Fantin-Latour (1836–1904) hauntingly summed up the end point and high point of a historical trajectory leading from progressive internalization to invisibility of knowledge (fig. 186). Meaning was grasped in isolation and retained as private property.[12] The dominion of texts over the bourgeoisie also entailed a domination of the page, scanned and probed from afar. Narrative, as Fantin-Latour's grave female sitters reveal, is antithetical to spontaneous, gestural speech just as print lacks the cross-cutting power and physical presence of the figural.[13] The identical shift from recognizing to listening occurred at the inception of psychoanalysis when Freud transformed Charcot's optical markers of hysteria into an archaeology of unimageable repressions.[14]

186

Henri Fantin-Latour

*The Reading*

1870

It was not that texts quantitatively replaced images. They supplanted them *qualitatively* for intellectuals as well as for middle-class audiences as the avenue for meaningful communication. Pierre-Auguste Renoir (1841–1919), the most rococo artist of the impressionist circle, demonstrated the extent to which pictorial designs had lost their cognitive quotient (fig. 187). In a portrait more revealing of social values than personal identity, his future wife Aline Charigot is shown at leisure, slumped in a chair perusing a fashion magazine. Significantly, the subject of the picture is not the pretty young model with open mouth gaping at advertisements for well-dressed *flâneurs*. Rather, it is the illustrated journal itself that is represented as producing idle spectators. Gone is the eighteenth-century interactive culture of conversation with its visual theater of passionate and informative attitudes (see figs. 1, 4). Renoir's dazzling and dizzying parade of modish commodities belongs to a world where you can choose what to watch. He offers a hallucinatory vision of bluish bits of graphic data reaching out and grabbing the attention of the viewer and enticing her to buy. Prophesying the media universe of cable channels hawking expensive goods, Renoir captured the demotion of images to effortlessly ingested and remotely controlled broadcasts.

The reification of print-based language as the master paradigm for all serious signification and the stereotyping of nonverbal expression as belonging to the impulse-ridden Unconscious has implications far beyond the hermetic disputes of contemporary academics. Howard Gardner has remarked on the overwhelming sense that our educational system has failed, or, worse, is fraudulent even in those cases where it seems to have succeeded.[15] Children, he maintains, need to synthesize various kinds of knowledge, including that stemming from craft skills traditionally anchored in artistic production.[16] Gardner's seminal distinction between rote or conventional performance and genuine disciplinary understanding is also apt. His concept of multiple intelligences (MI), including the competence in comprehending visual media,[17] is part of a larger theory of "situated learning." Further, pioneering neuroscientists have confirmed that the brain is an evolving organ. Not only in childhood but in adulthood it continues to build and destroy, strengthen and weaken connections depending on the amount of intellectual exercise the neurons and synapses receive.[18] This revolutionary notion that people learn more effectively when they are engaged in rich and meaningful projects was foreseen in the eighteenth-century conviction that sensory experience is essential for learning *throughout* life. Entertaining educational games enriched personal development by fostering intuitive, craft, symbolic, and notational types of knowing.

187

Pierre-Auguste Renoir

*Young Woman Looking at an Illustrated Magazine*

c. 1880

I argue that we are returning to the oral-visual culture of early modernism. To be sure, our image world is more heterogeneous, fragmented, indeterminate, and speeded up because of computers and robotic systems.[19] Indeed, it is difficult to imagine students in the manipulative and combinatorial "movieola period" of video or of electronic texts returning to scribal techniques. It is all the more important, then, to understand the role of visual analogues for abstract conceptions such as human development, cognition, memory, intelligence.[20] No longer preliterate, we are postliterate.[21] Yet, ironically, even within "postcolonial critical discourses" emphasizing the global importance of "hybridity" and the value of "alterity," the temporal linearity of texts serves as a model for a transnational countermodernity.[22] Might the patterns and shapes of cultures, their transformations and shifting relations, not be explored more effectively through randomization, animation, computer modeling, and morphing?

Similarly, an educational system that is heavily weighted toward linguistic and logical-quantitative modes of instruction and assessment continues to rule in the new image economy of interactive media, digital optic discs, synesthetics, "smart TV," and the electronic book. This fundamental and persistent contradiction is depressingly evident in the math phobia and image illiteracy encouraged by the customary aniconic practices and values of our schools. How can professors foster proficiency in quantitative reasoning or encourage technical literacy if they fail to represent knowledge in the dominant media of our time? To implement change, the National Science Foundation has been funding a calculus reform program. Significantly, the fear and loathing of many students for quantitative reasoning in any form is to be alleviated through such efforts (and others like it, for example the University of Chicago School Math Project) by emphasizing the pleasures of group work. They also rely on attractively illustrated textbooks and labor-saving calculators. The emphasis on an ingenious and varied patterning of learning activities, stressed in eighteenth-century mathematical recreations, seems to be voiced in the current call for flexibility in instructional formats.[23] The Enlightenment would have recognized itself in the revolutionary notion that students should not be treated as receptacles of knowledge but as interactive participants in multiple ways of knowing.

The discrepancy between a powerful and institutionalized discursive theory and a maligned, but ubiquitous, technological practice compelled me to examine the history of alternative attitudes toward visual education and instrumentalized entertainment. Further, I was struck by an additional paradox. Some art historians, in their haste to embrace structuralist and poststructuralist interpretive strategies as a way of improving our lowly status vis-à-vis the rest of the humanities and social sciences, attempt to turn the visual arts into a "language" whose grammar and syntax must be "read." At the moment when ocular expertise is most needed by scholars working in traditionally nonrepresentational fields such as mathematics, law, medicine, book history—all about to be transformed by the image revolution—is our profession jumping ship?

Similarly, a schematic and slogan-ridden language frequently triumphs over gesture to constitute the subject matter of contemporary painting. Not the magic but the poverty of the text shapes the "lingoism" and "nominalism" of a second-generation conceptualism.[24] By paying unreflective homage to the prestigious word, recent installments at the Whitney Biennial not only spurn the notion that art might be constitutive of the cognitive but deny that mental and manual skill are required to *make* it.

Neither voluntary subjugation to the disciplinary "other" nor a haphazard multiplication of "inter-," "cross-," or "trans-" cultural studies will erode entrenched stereotypes. Although sometimes electronically collapsed onscreen, the socially constructed distances between image and text, speaking and reading, applied technology and pure science, museum and university, will not be reduced until the epistemic hierarchies in which such antinomies remain grounded are exposed. To that end, this book asked how spatial patterns configured abstract information and in what unexpected ways they played a role in the European Enlightenment. This question led to the exploration of artful scientific performances as well as the identification of experimental artistic strategies. "Mathematical recreations," collaborative magic shows, staged experiments, learning machines, and natural history exhibitions were variously devised to teach old-regime audiences through pleasing pictures or graphic displays. Significantly, popular education was truly popular. Late modernism, with its disciplinary discontents, could benefit from the reenchantment of instruction on the threshold of the interconnective millennium.[25]

### *Looking Back*

At a time when the nonreading television generation is having an impact on the newspaper circulation in this country,[26] and when, ironically, classes leading to visual skills are among the first cut by school districts feeling a financial pinch, the moment seems perfect once again for communicating knowledgeably with images on the hypermedia "page." Pixels are the movable type of the future. The conviction that we are in the midst of a sweeping visual revolution led me to examine the closest historical parallel to the advent of digital electronic communication. Chapter 1 retrieved an early modern example of image/word synergism. Illustrated books of "scientific" games, actually of technological innovations cutting across all symbol systems, developed from the graphic conceits of Jesuit polymaths. The ingenious and recondite diversions of della Porta, Kircher, Schott, and Bonanni were transformed in the "rational" amusements of Ozanam, Guyot, Halle, Hooper, and Brewster. Such enlightening entertainments, unlike their prototypes, were intended for profane eyes.

Designed to fill that dangerous interval when the mind was released from its duties, these playful pretenses were the ancestors to today's "paper engineering." Innocent distractions—contrived at first for bored courtiers, then for restless children, their parents and tutors, and, finally, for the public at large—relied on gadgets, peep shows, and engaging slides. Like late twentieth-century users of Ron van der Meer's enchanting Art Pack,[27] early modern viewers delighted in gazing inside optical boxes (figs. 50, 66) or camera obscuras (fig. 112), peering through telescopes and microscopes (fig. 105), projecting phantoms on smoke (figs. 48, 58), manipulating pendulums (fig. 99), and activating steam engines (fig. 132). Hand-colored pull-out plates, cutouts, transparencies, and pricked prints indicate that these multinational and often serial editions were as much toys as books. Dexterity and sensory coordination were requisite for such do-it-yourself pastimes dedicated to profitably filling leisure hours.

As the occult and courtly values of the seventeenth century gave way before the more open, secular, and middle-class aspirations of the eighteenth, witty puzzles formulated for sophisticated virtuosi and aristocratic curieux were gradually supplanted by utilitarian pleasures with broader appeal. This perfecting and entertaining adult education was overtly constructed to combat a counterculture of fanatical enthusiasts and miracle-mongering operators.[28] The "oriental despot" symbolized unenlightenment. More generally, the "Eastern" tyrant and his dionysian retinue epitomized the perils of foreign ways, the seductiveness of alien, optically appealing mysteries (see figs. 65, 66). According to Boulanger, Robertson, and Salverte, the incantatory sights spun by ancient and modern sorcerers were manufactured to enslave an ignorant and idle populace.

Stereotypical scenes of ruthless Arabs portrayed them as holding the weak in erotic thrall (fig. 188). The haggling merchants of W. J. Muller's intrigue riddled and picturesquely decrepit Cairo *Carpet Bazaar* were depicted selling sumptuous, if dubious, commodities. Themes of barbaric combat, unrestrained sexual license, fairy-tale luxury, and tattered opulence flowed from the baroque into the mainstream of a nineteenth-century fictionalized orientalism (fig. 189).

Before dismissing hucksters, self-promoters, and con artists as relics of a distant and backward era, recall that gadgets, reputedly curing everything from baldness to cancer, and cases of academic fraud or flagrant plagiarism are reported daily in the international press. The rise of wizards, the reprise of magic cults, and the return to exhibitions of feudal hocus-pocus (*quigong*) have recently dismayed China's Communist Party.[29] Like the enlighteners, authorities bemoan the gullibility of spectators and the charlatanism of white-coated masters who, aided by a huge television screen, claim to harness the mystical forces of life and make coins melt, snakes crawl through nostrils, and blood ooze from pores.

Westerners have no cause for pride. Joey Skaggs, the American artist-prankster, has successfully been pulling hoaxes for the past 20 years to point up the credulousness and ethical irresponsibility of the media. His most notorious impostures include the *Fat Squad,* a commando-style diet service that physically restrains clients from overeating, the *Roach Hormone,* a miracle drug purportedly curing acne, arthritis, and menstrual cramps, and *Portafess,* a movable confessional booth set up outside the 1976 Democratic National Convention at Madison Square Garden. The *Brothel for Dogs* (1976), a fictitious canine bordello where animal owners paid $50 to have their male dogs sexually serviced by "Fifi," a pedigreed French poodle, or the mutt "Lady the Tramp," led the ASPCA to send out armed investigators. ABC subsequently produced a news broadcast that was nominated for an Emmy Award! Significantly, when the project was revealed as a hoax, the network refused to retract the story. Tricks and the intent to dupe the public also accompanied many supposedly serious moments in high modernism.[30] From the cerebral jokes of Marcel Duchamp to the gamesmanship of André Breton to the zany antics of Salvador Dalí, ironic playfulness undercut the grave process of making objects.

188

Eugène Delacroix

*The Abduction of Rebecca*

1846

189

William James Muller

*Carpet Bazaar, Cairo*

1843

Rational recreations, then, were created to counter irrational recreations. They reflected the Enlightenment belief that society might be perfected, that progress would lead to eventual utopia from which corruption had been eradicated. The "visible invisible" of chapter 2's title alludes to the literal and metaphorical juggling characterizing political, religious, industrial, and artistic "systems of imposture" denounced by eighteenth-century reformers. Dictionaries and encyclopedias of edifying entertainments taught the unwary consumer how to guard against fast moves. Quacks, mountebanks, and charlatans shared the page with gamblers, pickpockets, and counterfeiters. Labor-saving devices such as copying machines (fig. 86), measuring implements (fig. 90), silhouette tracers, and mechanical prints were included in the general indictment against fraudulent goods. Just as enlighteners combated the phantasmagoria of Mesmer and Cagliostro, neoclassical painters and theorists rose up against rococo artisans whose skill in *trompe-l'oeil,* they claimed, depended upon nonintellectual knack. Moreover, pilfering was inseparable from private enterprise and a growing market economy. Every trade had its distinctive cheats and illicit forms of remuneration. Eighteenth-century employers apparently were determined to make ancient customary rights criminal by having the former perquisites of craftsmen designated as stealing.[31]

Ernest Gellner has christened modern philosophy since Descartes a "comparative diabolics."[32] The *philosophes,* as well as quantifying natural philosophers, wanted to outwit the devil of the imagination as well as the demon of the senses. Logic was to pierce the tricks of the mind, language was to tear the veil of imagery, and theory was to unmask the idolatry of the laboratory's satanic setups. Chapter 3 exposed the compromising connections between conjuring and experimentation, between the fantastic theater of marvels and a demonstrating science. Unfortunately, the performances of empirics and empiricists were often disconcertingly alike. On one hand, Desaguliers, Whiston, Walker, Nollet, Rabiqueau, and Sigaud de La Fond were failed magicians because they believed in revealing the causes behind their special effects. On the other hand, as dexterous showmen turning their bodies into objects for scrutiny—especially during electric and magnetic trials—they were difficult to distinguish from stereotypical mountebanks. Folkloric scam artists, such as "Malpigiani," "Jerome Sharp" or "Crook-Finger'd Jack," deployed ingenious devices, concocted fake visual evidence, and fabricated pseudoevents impossible to replicate or verify.

The apparently all-too-easy slide of some scientists into data fudging produced a midcentury explosion of treatises. These manuals attempted to codify the rules for an incorruptible observation and a responsible "art of experiment," one impervious to the freakish lures of the sideshow. I argue that such tracts on *proper* know-how and "sincere" instrumentation helped to shape an intellectual climate conducive to a new kind of "honest" painting. What is conventionally termed *neoclassicism* is really a turn away from the supposedly passive "Catholic" or "southern" style of superstition to a "Protestant" and "northern" activist construction of personal and social identity. A spate of pictures and prints demonstrated, without recourse to illusionistic tricks, the manner whereby a multiplicity of things were actually done. Such works, I suggest, pictorialized the Enlightenment ideal of progress as tireless doing.

This international experimental art was fully the match of a pan-European artful science. Separating themselves from the strut, swagger, and swank depicted by Nattier, Raoux, and Fragonard, Chardin, Kauffmann, Mengs, David, and, on occasion, Reynolds and Gainsborough recorded cases of specific praxis. Accomplishments were embodied as technically precise performances. Incarnating a progressivist agenda, sitters actively *did* things rather than indolently pose or lounge around. As forms of visual education, demanding and demonstrating pictures trained their audiences to distinguish peacockery from creative energy.

Like the Abbé Nollet or Benjamin Martin, these informational painters constructed their publics by dramatically placing a variety of people and things into a variety of situations to illuminate their intercommunication. Making art was like conducting an experiment insofar as each yielded artificial phenomena. Neither the scientist nor the artist merely copied or imitated nature. Rather, in the manner of David's oath-taking militants, situations were contrived to force culturally invisible behavior or unseen events to manifest themselves (fig. 190). Cogent pictorial exemplification, then, could claim to be historically accurate, removed from courtly flattery and religious histrionics.

Greuze's visual *tableaux* also shifted attention away from the linear narrative to the pattern of social actions. Like Wright of Derby (fig. 76), he specifically alluded to well-known conjuring ploys that displaced the importance of what was done onto what was felt. The viewer was drawn through sympathy into a private drama that was simultaneously a public

190

Jacques-Louis David

*The Oath of the Horatii*

1786

performance (fig. 73). Like the strong points in a magic act, young women paused to stroke a bird (fig. 74) or smell a rose (fig. 75). A nubile laundress froze for an instant to entrap the spectator into completing an unspoken impulse (fig. 146). Silently filling in the inaudible blank was a libertine inversion of internalizing catechistic lessons. The constraints of decorum and etiquette were temporarily escaped, allowing the spectator to heed the emotional responses of his body.

Degas's industrialized disassembling and combinatorial strategies[33] were the culmination of this Enlightenment experimental art. Like Nollet's skilled operation of diverse machines (figs. 41, 47, 99), the later French artist's nuanced manipulation of corporeal motion[34] made hidden or overlooked human experience exhibit itself. His empirical studies of the heroism of modern life involved the visual analysis of an encyclopedic array of individual gestures and poses. Artificially thrust into odd but commonplace settings or caught in idiosyncratic but revealing situations, his characters were constrained either to display natural functions unnaturally or to exhibit normally concealed processes. By exposing the infinitely fine adjustments that go into trying on and settling into a closely fitting hat, Degas demonstrated the strangeness of even the most ordinary Parisian rituals (fig. 191). The deft manner in which a laundress lifted the recalcitrant front of a shirt with her index finger while ironing offers nonverbal proof of an elegant manual dexterity unexpected in such a lowly profession (fig. 192).[35]

Sensitive to every aspect of modern urban culture and commerce, the painter's eye was a finely adjusted registering and inscribing device. His custom of gauging incremental changes in himself and in any person he happened to be observing was reminiscent of Musschenbroek's notation that his hands tended to grow transparent when heated. Consider the complex ways in which the artist showed how the fleshy fingers of a New Orleans merchant became measurably obliterated when plunged into soft billows of cotton (fig. 193). Degas's learned execution, then, challenged Cabanal's *art pompier* sleights-of-hand. Perceptible networks of brushstrokes and conspicuous webs of pastel removed the master's facture from tricky maneuvering. His informed and informing manual skill, not mechanical legerdemain, was an enlightened style conveying his and others' craft.

191

Edgar Hilaire Germain Degas

*At the Milliner's*

1882

192

Edgar Hilaire Germain Degas

*A Woman Ironing*

1874

193

Edgar Hilaire Germain Degas

*The Cotton Merchants*

1873

On one hand, Degas subscribed to the Enlightenment's ethos of ceaseless experimentation and self-maximizing doing. Producing original and serial images was central to the linear rhetoric of unlimited progress. On the other hand, he realized the importance of constructing stable social identities that could achieve dignity within a recognizable cultural sphere. In this, he resembles contemporary ecologists who advocate not production but re-production. They urge all forms of design practice to visualize regenerating scenarios permitting a variable humanity and fragile phenomenal world to cohabit the planet.[36]

Eighteenth-century thinkers crystallized the modern dualistic attitude toward technology. Techniques were both the artificial instruments for attaining progress and the contaminated sources of an endless stream of polluting products. The tension between applied skill and theoretical learning is found in the notion that technology was merely what engineers automatically did as opposed to a programmatic basis for science. From the high vantage of a quantified natural philosophy, globes, orreries (fig. 55), pyrometers, barometers (fig. 47, 111), air pumps (fig. 76), electrical apparatuses (fig. 54), diving machines (fig. 77), and water mills (fig. 108) were revenue-producing commodities or, at best, mechanisms for solving limited practical problems. Leupold, Besson, Bion, Bélidor, Duchesne, Réaumur, Martin, and Ferguson—as promoters of philosophical entertainment and makers of a non-university-based natural philosophy—bore no resemblence to the illiterate Baconian craftsman. But the making, selling, and graphic dissemination of laboratory equipment as well as the deployment of auto-operated tools raised the specter of illicitly bypassing human labor. Then as now, work magically produced without bodily exertion aroused both suspicion and delight.

Vaucanson's self-propelled automata were the metallic counterparts to the earth imagined as an effortlessly unrolling and clocklike spectacle. Chapter 4 explored the growing dichotomy between an aesthetic, feminized contemplation of the pageant of the natural world and an observational, masculine scrutiny that sought to probe beneath the passing parade.[37] Fontenelle, Addison, and particularly the influential Abbé La Pluche separated the profundity of the arduous disciplinary gaze from the ceremony of pleasurable browsing. The guided tour of the universe as a succession of sanitized scenes collided with the close reading of the book of nature as an interpretable document.

Today, we remain the heirs to what Ada Louise Huxtable has termed "themed entertainment."[38] Although written about the preservationist movement and the dubious blending of old and new elements in architectural restorations, her designation is apt for all substitute realities, including those served up in museums and Disneylands. The custom of excerpting and permuting fragments of history and nature to create synthetic environments is, I argue, an outgrowth of stripping the surface of appearances and turning the "skin" of things into consumable goods. La Pluche was a master of illusion, marketing the false-front wonders of the universe. With the praiseworthy intention of enticing his aristocratic charges into admiring God's handiwork, he inadvertently pried apart sophisticated perception from the ocular shopping spree.

The pedagogical dilemmas surrounding the installation of natural history objects bear directly on the question of how any visual display conveys information and for whom. The two-sidedness of images, their capacity for dramatic falsification as well as for providing models for conceptualizing mental processes, continued to remain a burning issue for festival designers during the French Revolution. Faced with the paradox of conveying the superstition of past religious imagery through images, they puzzled over how not only to train the collective intelligence of a disparate public, but how to school the unruly senses. Gaudy floats and papier-maché monuments were signs of an invincible malaise leading to the eventual demand for an imageless spectacle without delusion.[39]

Just as the routines of philosophical entertainers were perceived to be uncomfortably close to juggling by the scientific establishment, so polymathic cabinets of rarities were taken to present only their owners' caprices. From the perspective of rational classifiers, collections accessible to everyone were not to be idiosyncratic and were obliged to shape the unified public they claimed to serve.[40] This major quandary colors, I think, all private and intimate eighteenth-century forms of performance caught in the transition from small to large sites where they received maximum exposure. Notably the rococo artisanal movement, with its cult of personal, irrational, and restless interior decoration,[41] was undercut by monumental, antiillusionistic, Salon painting conveying great ideas across grand and crowded spaces.

The fragile organic synthesis achieved by Daubenton at the *cabinet d'histoire naturelle* collapsed in the Rymsdycks' abstract procedure of excerpting minutiae. Looking ahead to the positivist historiographers of the nine-teenth century engaged in the anatomization of ever more finely dissected parts (fig. 194),[42] these self-consciously "scientific" draftsmen hoped their isolated pictorial documents might one day be fitted together into an overall pattern. Yet they had no way of envisioning such a totality except as an aggregate of additive details.

Disintegration and the radical hypostatization of particulars were charac-teristic also of romanticism. Marked by an alternating attraction for the circumstantial, contingent, and physical as well as for the universal, sym-bolic, and spiritual,[43] this complex period exacerbated the eighteenth-cen-tury tug-of-war between quantifiable singularities and sublime generali-ties.

### Looking Past

To conclude, I would like to suggest certain key, and still unexplored, areas where Enlightenment "tournaments of value"[44] between splitting and lumping things erupted into nineteenth-century visual culture. It seems clear that the concern with the individual as a conspicuous person-ality ideal,[45] surfacing at the time of the French Revolution in the writ-ings of Friedrich Schlegel, Fichte, and Schleiermacher, was an outgrowth of the extremes of eighteenth-century experimentation. The body tricks of the popular science demonstrator contained the seeds for the Jena Circle's impulse to open-ended self-exploration.[46] In both cases, an ab-sence of definition coexisted with a desire to meld with the force-filled and identity-giving cosmos. Cagliostro's adepts, radical Free Masons, ani-mal magnetists, Swedenborgians, Illuminists, and Rosicrucians continued an earlier generation's passionate attempt to fuse with a totality greater than themselves. Similarly, Vaucanson's high-performance automata estab-lished the impossible dream that the innovative scientist could engineer anything and that a utopian technology, in particular, could surmount any frailty still lurking within human biology.

Jerome McGann claimed that romantic poetry was marked by exaggerated kinds of displacement whereby pressing social concerns were resituated from an external to an internal or ideal realm.[47] As Schiller put it, beauty had priority over political freedom.[48] McGann's observation is pertinent also for the importance assigned to play in romantic aesthetic theory. The popular recreation industry of the eighteenth century deployed games

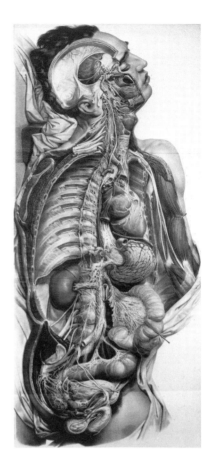

194

Jean-Baptiste-Marc Bourgery

*General Anatomico-Physiological*

*Summary of the Sympathetic Nerves*

1844

from *Anatomie descriptive de physiologique*

both as educational tools, or intellectual exercises, and as formal diversions, expressing the ingenuity of their makers. The art of demonstration, then, oscillated between serious culture formation and the playful cultivation of the imaginative self.

This new type of aestheticized human subjectivity[49] was tenuously embodied in the substanceless and autonomous arabesque.[50] On one hand, bodiless ornament, like musical rhythms and interlacing phantasms, mimicked the conjuror's *tours de passe-passe*. On the other hand, Schiller argued that the play impulse did not permit the artist to weave egotistical abstractions. The creative imagination had, instead, a responsibility to transform the art of living through a community of scientists, artisans, and scholars going about their ordinary affairs, but with a new attitude deepened by feeling.[51] This kind of knowing by subtle modes of sensuous perception[52] owed to the aesthetics of Enlightenment entertainment. Fantasizing, like the optical cabinets that were its stimulus, wavered between significance and absurdity, useful learning and empty decoration.[53]

Caspar David Friedrich's *Rückenfigur* (see fig. 185), or archetypal individual seen from the back,[54] incarnated this dualism endemic to the romantic conception of play. The *Woman at a Window* is both a mysterious apparition, conjured with no visible sign of how she came into that deserted room, and a concrete presence, whose intense gaze forces ours to follow hers into the landscape.[55] In this picture, ephemeral water and light possess the ontological lightness of the arabesque. This weightless line symbolized the fragility of the encounter between a transitory psyche and a vanishing external flow. Elinor Shaffer has aptly described the melancholy of the romantic system,[56] a Platonic cave where dim figures of potentiality sharpen for an instant into *trompe-l'oeil* images before lapsing again into obscurity. Friedrich wanted to reestablish a connection between filmic appearances, vying for attention in an evanescent environment, and a fleeting human consciousness. Resembling La Pluche's spectacle gazers, however, his viewers exfoliated the visible world, decredentializing it in the quest for transcendental symbols and in the pursuit of stable identities for themselves.[57]

Not enough has been made of the fact that in attempting to harmonize a horizontally processing nature with their vertical yearnings, the romantics reinforced anthropocentric hierarchy. The thinking and daydreaming subject remained an intruder on the material scene. In this they were heirs to Rousseau who had claimed that sensations were the sole reality.[58] But the *grand solitaire*'s reverie was generated by the gentle merger of his dissolving ego with insubstantial phenomena, whereas Friedrich's onlookers possessed no analogous elasticity of mind. Play, however, achieved artificially and negatively what could no longer be accomplished naturally and positively. The liberating procedures of combinatory wit and deflating irony chemically dissolved dialectical oppositions.[59] Analogously, eighteenth-century experiments as deconstructive acts constantly unfixed the systems raised by analytical philosophy. In the process, irreducibly human substances resisting easy categorization were precipitated out.

The romantic virtuoso epitomized the magnified and unclassifiable self-as-art. Descendants of scientific spellbinders, these blatant mesmerizers also turned their bodies into a dramatic experience by exposing vulnerable flesh as if it were invulnerable (see figs. 104, 128, 131). Such overwrought communication of the "unsounded self" of the actor[60] to an emotionally receptive audience shaped a melodramatic strain in romantic painting. Francis Danby's (1793–1861) disorderly *Precipice* relied on visual hyperbole and a welter of material things to impress the spectator (fig. 195), who was stunned by an excess of craggy rocks, swirling mists, and billowing waterfalls. Like the interrupted and disjointed phrases occurring in exaggerated stage dialogue,[61] sublimely high mountains and improbably deep chasms were arbitrarily hacked off at the bottom and top of the picture to intensify the psychological impact. The defenselessness of the painting's partially clad Alpine hunter, suggestively outstretched in a threatening and crudely painted setting,[62] is reminiscent of the convulsive sexuality exuded by eighteenth-century electrifiers.

Recklessness was precariously close to the distorted look of caricature. Eugène Delacroix (1798–1863) believed that such extreme deformations captured the grotesque aspect of modernity.[63] Like Byron or Nodier, he drew on the violence implicit in the Voltairean invective against superstition and thematized it into fantastic images of combat between exotically different races.[64] In his many scenes of Arab strife, the Enlightenment binary opposition between Western reason and "Asiatic" credulity was metamorphosed into magical turmoil. The *Abduction of Rebecca* (see fig. 188) permitted an alien chaos to coalesce almost mythically for an instant. Ideal order and stillness were wrested from a riot of colors and contradictory parts. The beautiful and helpless Rebecca, caged by an interlace of swarthy arms, is the ironic and melodramatic inversion

195

Francis Danby

*The Precipice*

19th century

of Fragonard's blindfolded and tickled girl-toy (see fig. 133). Sir Walter Scott's heroine, as the entrapped victim of tyrannical rape, also forms the ironic female counterpart to Rabiqueau's willed bondage, his voluntary submissiveness to cosmic violation (see fig. 128).

Many of Delacroix's major compositions during the Restoration—the *Massacre of Chios* (1824), *Greece Expiring on the Ruins of Missolonghi* (1827), and *The Death of Sardanapalus* (1828)—were both moral fables directed against autocratic absolutism and paeans to its alluring charms.[65] Consequently, they advanced Enlightenment pedagogics bent on extirpating "Eastern" barbarity while ironically extolling supernaturalism. By juxtaposing revolt with exorcism, the beautiful with the ugly, the divine with the diabolical, these works realigned the eighteenth-century preoccupation with corruption into an obsession with the fallenness of mankind. According to Baudelaire's conception of caricature, the absolute came into being only through the partial and imperfect perspective of a grotesquely distorted humanity.[66] Delacroix similarly held out the hope of redemption in incongruous images where virtuous resistance was pitted unequally against nightmarish forces. Modern allegories of violence, concupiscence, and Molochistic sacrifice, these victim-strewn paintings challenged and celebrated arbitrary power by displaying it condensed in an awe-inspiring foreign oppressor.

Baron Antoine-Jean Gros (1771–1835), whom Delacroix admired for his sinister canvases showing real cadavers rotting in the muck, orientalized Napoleon's career.[67] Throughout the Empire period, his exotic spectacles of Near Eastern campaigns demonstrated the fallacies of military victory. The progressivist impulse of the Enlightenment soured into unchecked self-aggrandizement, and the regressive striving for infinite potency destroyed the liberty of others.[68] Gros's vast pictures of the satrap-conqueror, destined for exhibition at the official Salon, were subtly subversive of the power and authority of Napoleon whom they ostensibly glorified. An equivocal image of absolute rule emerged in the nexus of individual insatiability and horrific mass excess. The thaumaturgic emperor, visiting the bloody battlefield of Eylau at dawn on February 9, 1807, was silently mocked by the brutal pile of dead soldiers he could not conjure back into existence (fig. 196).

Drawing upon the aggression inherent in Enlightenment dichotomies, the French painter externalized this violence by the raw juxtaposition of

196

Antoine-Jean Gros

*Napoleon on the Battlefield of Eylau*

1807

ostensive panegyric with illuminating critique. Recalling Robertson's and Salverte's debunking of fanatical priestcraft, the artist similarly unmasked the ruler's drive to dominate through the false promise of redemption. Worshipful Russian and Prussian warriors were depicted as submissive to the emperor's delusion that he could draw down the power of heaven and cure the wounded.[69] In fact, actual caring for the sick was being carried out by seven military doctors shown valiantly bandaging and operating in the muddy chaos.

To further highlight the distance between lie and truth, the myth of a kingly healing touch and the disastrous carnage, Gros had the modern magus crystallize mysteriously in the midst of a gray and frozen panorama. Napoleon appeared out of nowhere, like a ghost at a phantasmagoric seance. Such allusions to the showy despot as gulling mountebank were underscored in the failed pseudoexperiment documented in the foreground. The mound of unresurrected corpses offered rational evidence for the impotence of irrational wizardry.

Edouard Manet (1832–1883) revealed his debt to the anti-Catholic rhetoric of the Enlightenment by exposing the "imposture" of divine intervention in human history more explicitly. His *Dead Christ with Angels* exhibited the Son of God like a natural history specimen propped up in a museum case. Following the positivistic analysis of the orientalizing philologist Ernest Renan,[70] the venerable God of Christianity became scientized into a gangrenous mortal. Unlike the hieratic immutability and corporeal idealism favored in ultramontanist representations, Manet turned sacred precinct into barren laboratory for the clinical observation of clotting blood and festering stigmata (fig. 197). Stripped of secrecy and bereft of the hidden levers associated with necromantic ploys, the supreme conjuror was compelled to confront his spectators. Like Pinetti in performance, the artist asked the audience to see and decide (see fig. 72).

For a self-consciously naturalistic and antisuperstitious age, mired in sclerotic dualism,[71] the prophet who claimed he would rise after three days in the tomb proved false. The rigid pose resembled not only the stuffed museum mount but the mechanical toy. Remember that Courbet's figures had been routinely compared to wooden dolls and childish automata in midcentury Parisian journals of art caricature.[72] If Christ's half-standing, half-sitting stance mimicked the android's stiffness, his decomposing flesh was that of the delinquent.[73] A wholly material phy-

197

Edouard Manet

*The Dead Christ with Angels*

1864

*Conclusion*

sique and physiognomy bore the criminal marks of crucifixion. This agnostic comparative anatomy, like Nicolas-Henri Jacob's draftsmanly disintegration of the cadaver into tangible "vérités palpables," claimed to set facts "naturally" before the eyes.[74] The plates for Jean-Baptiste-Marc Bourgery's *Traité complet de l'anatomie de l'homme comprenant la médicine* (1831–1854) were an instantiation of positivist hermeneutics, a leveling linguistics of the body. Recalling Nehemiah Grew's unpretentious specimens (see fig. 170), the miracle of the nervous system was transformed into materialistic prose (fig. 194). Yet this pitiless literalness, like Manet's debunking rendering, paradoxically forced face, torso, and limbs to grimace.

But it was the romantics as the heirs of the Enlightenment, who first made possible Courbet's and Manet's corporealization of the representational field.[75] Their insertion of teratological, pathological, or distorted elements into painting increasingly turned the body into a gymnastic spectacle. As Hazlitt remarked, the precision exercises of mechanical performers stood in opposition to the "ethereal, evanescent, refined, sublime part of art" that filtered nature through the medium of sentiment. Mastery over material execution, embodied in the overt manipulation of paint and pigment, was also the antithesis of the slow cerebration of the intellectual. The essayist compared the agile performance of the rope dancer, Richer, to the art of the "professor," Reynolds. The graceful subjects of the latter's portraits were not objects of sight but evoked feeling, and thus appealed to a "finer sense" (see figs. 124, 125). Nevertheless Hazlitt had to admit that if the famous acrobat had left as many gaps and botches in his performance as the celebrated President of the Royal Academy had included in his, the circus star would have broken his neck.[76]

The British writer's view of the artisan as being without theoretical concepts and the artist as cheating in the production of well-crafted work points to the potential for deception in virtuosic practice. The desire to achieve a reputation for dazzling genius, as Norman Fruman has convincingly demonstrated, caused Coleridge to engage in extended plagiarism from Schelling, Akenside, Gray, and Wordsworth. The romantic ideology of prophetic originality and the desire to create a totality from partial knowledge apparently drove him to posturing, insincerity, and the lifting of uncredited citations from which he pieced together his major poems.[77] Harking back to a baroque aesthetics based on prestige and visibility,[78]

Coleridge radically realigned that earlier era's system of ostentatious physical expenditure into the show of extravagant mental power.

Driven to seek reconciliations and to reduce all philosophical conflicts to harmony,[79] the English poet and critic felt constantly compelled to generate stunning intellectual inventions out of the ruins of a past poetics. Paradoxically, the desire to transcend the supposed mechanism and materialism of the eighteenth century as well as its political corruption in the name of a higher, invisible ideal led precisely to that charlatanism Enlightenment educators had warned against.

If it is true that present-day neoromantic artists focus more and more on themselves and reach out less and less to their audiences,[80] then nineteenth-century developments offer no consolation. Rather, it was in the eighteenth century's demonstration of pleasurable learning that aspects of personal experience were put at the service of a public beyond the borders of the narcissistic self. The activity of attractively making knowledge visible not only kept the performer going, but engaged the viewer to constructively play along.

We are currently witnessing an exponential increase in competing computer domains and, with this electronic explosion, a proliferation of specialized, and potentially isolating, visual expertises. One can no longer speak of a single visual culture. Scanners, monitors, and high-quality laser printers constitute the late twentieth-century mythic "East" of incantatory sights and digital tricks. As more and more personal computers can read in and alter images from the printed page, or are hooked to CD-ROM readers and disks filled with thousands of graphics, the likelihood of widespread theft, manipulation, and promiscuous and unauthorized copying becomes greater. The availability of on-line journals and the quickness of electronic publishing and distribution raise troubling questions not only about the circumvention of peer review but about the likelihood of people stealing or tampering with data in all formats.

Nontheoretical spectacle and noncognitive and unethical sleight-of-hand are the twin poles of our orientalized technology. Automatic simulations and mechanized knack remain the staples of that media-"Asia" of credulity generally claimed to be holding a global network of passive viewers in thrall. Those of us with knowledge about the techniques for making and understanding images and their constructive, *cognitive* role throughout history had better speak up now or be content to vanish into disciplinary extinction. The notion that the typeset text forms the rule and highest principle of interpretation must finally be put to rest for a democratic hermeneutics of pattern recognition and visual design to be created. As mass media become increasingly private and particularized, all forms of graphic display will have to be reassociated with common rituals and public concerns. This was the lesson of the Enlightenment. As the Reenlightenment of the twenty-first century glints on the horizon, the transit of information rushes forward into the past.

*1*

THE MIND'S RELEASE

1. Ludmilla Jordanova, "Objects of Knowledge: A Historical Perspective on Museums," in *The New Museology,* ed. Peter Vergo (London: Reaktion Books, 1989), p. 23.

2. Emily Martin, "Science and Women's Bodies: Forms of Anthropological Knowledge," in *Body/Politics: Women and the Discourses of Science,* ed. Mary Jacobin, Evelyn Fox Keller, and Sally Shuttleworth (New York and London: Routledge, 1990), pp. 1, 70.

3. Chandra Mukerji, *From Graven Images: Patterns of Modern Materialism* (New York: Columbia University Press, 1983), pp. 12, 132–133.

4. Colin Campbell, *The Romantic Ethic and the Spirit of Modern Consumerism* (Oxford: Basil Blackwell, 1987), p. 38.

5. See Neil Postman, *Amusing Ourselves to Death: Public Discourse in the Age of Show Business* (New York: Elizabeth Siffon Books, Penguin Books, 1989), p. 146; and Shanto Iyengar, *Is Anyone Responsible? How Television Frames Political Issues* (Chicago and London: University of Chicago Press, 1991), pp. 11–14.

6. For overviews of this change, see Jack Goody, *The Logic of Writing and the Organization of Society* (Cambridge:

Cambridge University Press, 1986), p. 20; and Alvin Kernan, *Printing Technology, Letters, and Samuel Johnson* (Princeton: Princeton University Press, 1987), pp. 4, 48–51.

7. Bruno Latour, *Science in Action: How to Follow Scientists and Engineers through Society* (Milton Keynes: Open University Press, 1987), p. 21.

8. For the importance of speech and hieroglyphics in "primitive" times, see Antoine Court de Gebelin, *Monde primitif, analysé et comparé avec le monde moderne; considéré dans son génie allégorique et dans les allégories aux qu'elles conduisit ce génie,* 9 vols. (Paris: Chez l'Auteur, 1773–1782), VIII, xii.

9. On the importance of Pyrrhonism as a conceptual framework, see Terence Allan Hoagwood, "The Negative Dialectic of Byron's Skepticism," *European Romantic Review,* 3 (Summer 1992), 21–40.

10. See Fernando Poyatos, "New Perspectives for an Integrative Research of Nonverbal Systems," in *Nonverbal Communication Today: Current Issues,* ed. Mary Ritchie Key (Berlin, New York, Amsterdam: Mouton Publishers, 1982), p. 121; and Howard Gardner, *Frames of Mind: The Theory of Multiple Intelligences* (New York: Basic Books, 1983), p. 173.

11. Mihaly Csikszentmihalyi, *Flow: The Psychology of Optimal Experience* (New York: Harper and Row, 1990), p. 26; and Csikszentmihalyi and Eugene Rochberg-Halton, *The Meaning of Things Domestic Symbols and the Self* (Cambridge: Cambridge University Press, 1981), pp. 193–194. On meaning-bearing objects, also see Krzystof Pomian, *Collectors and Curiosities: Paris and Venice, 1500–1800* (Cambridge: Polity Press, 1990), p. 5.

12. Thomas De Quincey, "Conversation" [1847], in *Selected Essays on Rhetoric,* ed. Frederick Burwick (Carbondale and Edwardsville: Southern Illinois University Press, 1967), p. 271.

13. De Quincey, "Rhetoric" [1828], in *Selected Essays,* p. 97.

14. For an excellent review of English seventeenth-century polemics, see David Vincent, *Literacy and Popular Culture: England 1750–1914* (Cambridge: Cambridge University Press, 1989), pp. 5–6.

15. See the famous case of the Catholic vicar of Etrepigny whose last will and testament was a scathing indictment of the "charlatanism" of the Church: Jean Meslier, *Superstition in All Ages by . . . a Roman Catholic Priest, or . . . Common Sense* [1762], trans. Anna Koop (New York: Truth Seeker Company, 1950), pp. 51, 310ff. On the libertine revolt against

Christianity, even within Catholicism, see Peter Wagner, "Anti-Catholic Erotica in Eighteenth-Century England," in *Erotica and the Enlightenment,* ed. Peter Wagner (Frankfurt-am-Main: Peter Lang, 1991), pp. 167–168.

16. See Julius S. Held, *Rembrandt Studies* (Princeton: Princeton University Press, 1991), p. 170; and Gary Schwartz, *Rembrandt: His Life, His Paintings* (New York: Viking, 1985), pp. 217–219. On the Protestant need to have someone speak for the Scriptures, see Joel C. Weinsheimer, *Eighteenth-Century Hermeneutics: Philosophy of Interpretation from Locke to Burke* (New Haven and London: Yale University Press, 1993), p. 9.

17. For the development of hermeneutics from Spinoza forward, see Vassilis Lambropoulos, *The Rise of Eurocentrism: Anatomy of Interpretation* (Princeton: Princeton University Press, 1993), pp. 23–29.

18. Paul Monod, "Painters and Party Politics in England, 1714–1760," *Eighteenth-Century Studies,* 26 (Spring 1993), 371–375.

19. For the importance of the Book of Tobit, see Held, *Rembrandt Studies,* pp. 118–120.

20. Zirka Zaremba Filipczak, "'A Time Fertile in Miracles': Miraculous Events in and through Art," in *The*

*Age of the Marvelous,* exh. cat., ed. Joy Kenseth (Hanover, N.H.: Hood Museum of Art, Dartmouth College, 1991), pp. 193–203.

21. On the connection between monotheism, rationalism, and the drive toward intellectual unity, see Walter Pape, "Heiliges Wort und weltlicher Rechenpfennig. Zur Entwicklung des Sprachauffassung im 17. Jahrhundert (Jacob Böhme, Athanasius Kircher, Leibniz)," in *Religion und Religiosität im Zeitalter des Barock. Wolfenbütteler Arbeiten zur Barockforschung* (Wiesbaden: Harrassowitz, 1993), pp. 12–13. For the aesthetic ramifications already evident in early Christian debates, see Gilles Sauron, "Les Monstres au coeur des conflits esthétiques à Rome au 1er siècle avant Jésus Christ," *Revue de l'Art,* 90 (1990), 37–42.

22. E. S. Shaffer, *'Kubla Khan' and the Fall of Jerusalem: The Mythological School in Biblical Criticism and Secular Literature* (Cambridge: Cambridge University Press, 1975), pp. 36, 194. For the prehistory of atheism, see Alan Charles Kors, *Atheism in France, 1650–1729.* Vol. I: *The Orthodox Sources of Disbelief* (Princeton: Princeton University Press, 1990).

23. See my *Body Criticism: Imaging the Unseen in Enlightenment Art and Medicine* (Cambridge and London: MIT Press, 1991), pp. 362–377.

24. On the European creation of the Orient, see Edward Said, *Orientalism* (New York: Vintage Books, 1978), p. 5. Significantly, he does not mention Boulanger.

25. Nicolas-Antoine Boulanger, *The Origin of Despotism in the Oriental and Other Empires of Africa, Europe, and America* (Amsterdam: n.p., 1764), pp. 3–13. Also see the excellent intellectual biography by Paul Sadrin, *Nicolas-Antoine Boulanger (1722–1759), ou avant nous le déluge* (Oxford: The Voltaire Foundation, 1986), pp. 3–5.

26. Barbara Maria Stafford, *Symbol and Myth: Humbert de Superville's Essay on Absolute Signs in Art* (Cranbury, N.J.: Associated University Presses, 1979), pp. 154–175.

27. Ibid., pp. 59–62. Boulanger cites Jacques Gaffarel, *Curiositez inouyes, sur la sculpture talismanique des persans, horoscope des patriarches. Et lectures des estoilles* ([Paris]: n.p., 1637), pp. 282–287.

28. Nicolas-Antoine Boulanger, *L'Antiquité dévoilée par ses usages; ou examen critique des principales opinions, cérémonies, & institutions religieuses & politiques des différens peuples de la terre,* 3 vols. (Amsterdam: Chez Marc Michel Rey, 1766), III, 342–348

29. On Montesquieu, see Carol Houlihan Flynn, *The Body in Swift and Defoe* (Cambridge: Cambridge University Press, 1990), p. 53.

30. Nicolas-Antoine Boulanger, *Le Christianisme dévoilé, ou examen des principes et des effects de la religion chrétienne* (London: n.p., 1756), pp. 121–124. There is some question as to both authorship and date of this volume. Sadrin, however, still leans toward Boulanger rather than d'Holbach as its author. It may not have been published until 1766 as a companion to the *Antiquité dévoilée*. See Sadrin, *Boulanger*, p. 142. On the need to reform the Church as perceived by eighteenth-century pontiffs, see the exemplary study by Christopher M. S. Johns, *Papal Art and Cultural Politics: Rome in the Age of Clement XI* (Cambridge: Cambridge University Press, 1993), p. 4.

31. Reginald Scot, *The Discovery of Witchcraft: Proving That the Compacts and Contracts of Witches with Devils and All Infernal Spirits or Familiars, Are But Erroneous Novelties and Imaginary Conceptions . . . and the Knavery of Juglers, Conjurers, Charmers, Soothsayers, Figure-Casters, Dreamers, Alchymists and Philterers; with Many Other Things That Have Long Lain Hidden, Fully Opened and Deciphered* (3d ed., London: Printed for A. Clark, 1665), The Epistle Dedicatory.

32. On the denunciation of the Jesuit collegiate system, see L. W. B. Brockliss, *French Higher Education in the Seventeenth and Eighteenth Century: A*

*Cultural History* (Oxford: Clarendon Press, 1987), pp. 171–174; and Charles R. Bailey, "Secondary Education," in *Facets of Education in the Eighteenth Century,* ed. James A. Leith (Oxford: The Voltaire Foundation, 1977), p. 108.

33. Christiane Klapisch-Zuber, *Women, Family, and Ritual in Renaissance Italy,* trans. Lydia Cochrane (Chicago: University of Chicago Press, 1985), pp. 310–312, 321.

34. Anthony Grafton, *Defenders of the Text: The Traditions of Scholarship in an Age of Science, 1450–1800* (Cambridge and London: Harvard University Press, 1991), pp. 230–231.

35. François Hemsterhuis, *Lettre sur la sculpture,* in *Oeuvres philosophiques,* 2 vols. (Paris: De l'Imprimerie de H. J. Jansen, 1792), I, 30–38.

36. Ibid., pp. 6–10. On the visual psychology of attraction and repulsion, also see his *Lettre sur les désirs,* in *Oeuvres,* I, 63–67, 73, 78.

37. Jennifer Montagu, *Roman Baroque Sculpture: The Industry of Art* (New Haven and London: Yale University Press, 1989), pp. 173–175.

38. Janis A. Tomlinson, *Francisco Goya: The Tapestry Cartoons and Early Career at the Court of Madrid* (Cambridge: Cambridge University Press, 1989), pp. 154–156, 197–200.

39. See, for example, the bilingual manual issued for the Barefoot Car-

melites of Bavaria, the *Ichnographia Triplicis ad Deum Tri-Unum Mysticae Viae, Purgativae, Illuminativae Unitivae: Splendoribus Sanctorum* (Augsburg: Ignatius Verhelst, 1779), pls. 5, 64, 71. For the derision of "superstitious delusion," see David Hume, *Of Miracles* [1748], intro by Anthony Flew (La Salle, Ill: Open Court, 1987), p. 25.

40. Eusèbe Salverte, *Des sciences occultes, ou essai sur la magie, les prodiges, et les miracles,* 2 vols. (Paris: Sédillot, Libraire-Editeur, 1829), I, 180, 185–186. On Goya's lifelong program of reform, see Priscilla E. Muller, *Goya's "Black" Paintings: Truth and Reason in Light and Liberty* (New York: The Hispanic Society of America, 1984).

41. E.-G. Robertson, *Mémoires récréatifs scientifiques et anecdotiques,* 2 vols. (Paris: De l'Imprimerie de Rignoux, 1831), I, 83, 165–167.

42. See Roger Chartier, "The Practical Impact of Writing," in *A History of Private Life,* vol. III: *Passions of the Renaissance,* ed. Roger Chartier, trans. Arthur Goldhammer (Cambridge and London: Belknap Press of Harvard University Press, 1989), 117–125, 130–132; and Campbell, *Romantic Ethic,* p. 31.

43. James van Horn Melton, *Absolutism and the Eighteenth-Century Origins of Compulsory Schooling in Prussia and Austria* (Cambridge: Cambridge University Press, 1988), pp. 22–23, 28–31. More generally on the Protestant desire for literacy, see J. Paul Hunter, *Before Novels: The Cultural Contexts of*

*Eighteenth-Century English Fiction* (New York and London: W. W. Norton, 1990), pp. 83–84.

44. Joseph Leo Koerner, *Caspar David Friedrich and the Subject of Landscape* (London: Reaktion Books, 1990), p. 77.

45. On the Calvinist demands made of the art object, see Donald Davie, *A Gathered Church: The Literature of the English Dissenting Interest, 1700–1930* (New York: Oxford University Press, 1978), pp. 25–26.

46. Jacques-Antoine Dulaure, *Des cultes qui ont précédé et amené l'idolatrie ou l'adoration des figures humaines* (Paris: Chez Fournier Frères, 1805), pp. 18–20. Also see William Pietz, "The Problem of the Fetish, II: The Origin of the Fetish," *Res,* 13 (Spring 1987), 28–30.

47. Jean-Jacques Rousseau, *Emile or On Education,* trans. Allan Bloom (New York: Basic Books, 1979), p. 33.

48. Augustin Roux, *Nouvelle ency-clopédie portatif, ou tableau général des connoissances humaines; ouvrage recueilli des meilleurs auteurs, dans lequel on en-treprend de donner une idée exacte des sci-ences les plus utiles, & de les mettre à la portée du plus grand nombre des lecteurs,* 2 vols. (Paris: Chez Vincent, Imprimeur-Libraire, 1766), I, vii–viii.

49. See Lloyd de Mause, *The History of Childhood* (New York: The Psycho-history Press, 1974), chaps. 1, 6.

50. George Turnbull, *A Treatise on An-cient Painting, Containing Observations on the Rise, Progress, and Decline of That Art amongst the Greeks and Romans: The High Opinion Which the Great Men of Antiquity Had of It, Its Connexion with Poetry and Philosophy; and the Use That May Be Made of It in Education* (Lon-don: Printed for the Author, 1740), p. viii.

51. George Turnbull, *Observations upon a Liberal Education, in All Its Branches: Containing the Substance of What Hath Been Said upon That Important Subject by the Best Writers Ancient or Modern: With Many New Remarks Interspersed: Designed for the Assistance of Young Gen-tlemen, Who Having Made Some Progress in Useful Sciences Are Desirous of Mak-ing Further Improvements* (London: Printed for A. Millar, 1742), pp. 48–49. On the empirical and inductive nature of Scottish secondary educa-tion, see Roger L. Emerson, "Scottish Universities in the Eighteenth Cen-tury, 1690–1800," in Leith, *Facets,* pp. 457–461.

52. Csikszentmihalyi, *Flow,* p. 73.

53. For the connection between a mannered style and ludic or ornamen-tal entertainment, see Giancarlo Maiorino, *The Portrait of Eccentricity: Arcimboldo and the Mannerist Grotesque* (University Park and London: Pennsyl-vania State University Press, 1991), p. 58.

54. [Henri-Gabriel Duchesne], *Notice historique sur la vie et les ouvrages de J. B. della Porta, gentilhomme napolitaine* (Paris: Chez Poignée Imprimeur, An IX), pp. 3–7, 211.

55. Ibid., p. 14.

56. [Giovanni Battista] della Porta, *Natural Magick: In XX Bookes by . . . a Neapolitaine* (London: Printed for Thomas Young and Samuel Speed, 1658), preface.

57. William Leybourne, *Pleasure with Profit: Consisting of Recreations of Diverse Kinds, viz. . . . . Published to Recreate In-genious Spirits; and to Induce Them to Make Farther Scrutiny into These (and the Like) Sublime Sciences; and to Divert Them from the Following Such Vices, to Which Youth (in This Age) Are So Much Inclined* (London: Printed for Richard Baldwin, 1694), p. ii.

58. Campbell, *Romantic Ethic,* p. 25. On the rise of educational games in the home, see Lawrence Stone, *The Family, Sex, and Marriage* (New York: Harper Colophon Books, 1977), pp. 258, 274.

59. Henry van Etten, *Mathematicall Recreations, or a Collection of Sundrie Excellent Problemes out of Ancient and Modern Philosophers Both Useful and Recreative* (London: Printed for Rich-ard Hawkins, 1677), title page.

60. Trevor H. Hall, *Mathematicall Rec-reations: An Exercise in Seventeenth-Century Bibliography* (Leeds: W. S. Maney and Son, 1969), pp. 3–7, 21–23.

61. Van Etten, *Mathematicall Recreations*, pp. 268–272.

62. Bernard Forest de Bélidor, *Le Bombardier françois, ou nouvelle méthode de jetter les bombes avec précision* (Amsterdam: Aux dépens de la Compagnie, 1734), pp. 312–313, 327–330, 359–360.

63. Hall, *Mathematicall Recreations*, p. 23.

64. Van Etten, *Mathematicall Recreations*, preface.

65. Henry van Etten, *Mathematicall Recreations. Or a Collection of Sundrie Problems Extracted out of the Ancient and Modern Philosophers, as Secrets in Nature, and Experiments in Arithmeticke, Geometrie, Cosmographie, Navigation, Musicke, Opticks, Architecture, Staticke, Machanicks, Chimistrie, Waterworkes, Fireworkes, etc. Not Vulgarly Made Manifest until This Time. Fit for Scholars, Students, and Gentlemen, That Desire to Know the Philosophical Cause of Many Admirable Conclusions* (London: Printed by T. Cotes for Richard Hawkins, 1633), dedication.

66. Claude-Gaspar Bachet, Sieur de Méziriac, *Problèmes plaisans et delectables, qui se font par les nombres: partie recueillis de divers autheurs, & inventez de nouveau avec leur demonstration* (Lyons: Chez Pierre Rigaud, 1611), preface.

67. Thomas Johnson, *A New Booke of New Conceits, with a Number of Novelties Annexed Thereunto. Thereof Some Be Profitable, Some Necessary, Some Strange, None Hurtful, and All Delectable* (London: Printed by E. A. for Edward Wright and Cuthbert Wright, 1630), p. 3.

68. Nicolas Hunt, *Newe Recreations. Or the Mindes Release and Solacing. In a Rare and Exquisite Invention for the Exercising of Acute Wits and Industrious Dispositions. Replenished with Mysteries, Secrets, and Rareties, Both Arithmeticall and Mathematicall; Not Formerly Discovered by Any. {And} Judiciary Exercises and Practicall Conclusions* (London: Printed by Aug. Math. for Luke Faune, 1631), The Epistle Dedicatorie.

69. Kenseth, *Age of the Marvelous*, p. 244.

70. Barbara Maria Stafford, "'Fantastic Images' or the Tradition of 'Appearances' Meant to Be Seen in the Dark," in *Aesthetic Illusion*, ed. Frederick Burwick and Walter Pape (Berlin and New York: Walter de Gruyter, 1990), pp. 158–182. Also see Gustav René Hocke, *Die Welt als Labyrinth. Manierismus in der europäischen Kunst und Literatur*, ed. Curt Grützmacher (rev. ed.; Reinbek bei Hamburg: Rowohlt, 1987), p. 152.

71. Kaspar Schott, *Mechanica Hydraulica-Pneumatica, qua Praeterquam quod Aquei Elementi Natura, Proprietas, Vis Matrix, atque Occultus cum Aëre Conflictus, a Primus Fundamentis Demonstratur, Omnis quoque Generis Experimenta Hydraulico-Pneumatica . . . Opus Bipartem* (Frankfurt: Joannis Godfridi Schönwetteri, 1656), pp. 416–419. See Duchesne, *Notice historique*, pp. 19–20, for the eighteenth-century perception of Kircher's importance.

72. Kaspar Schott, *Joco-Seriorum Naturae et Artis, sive Magiae Naturalis Centuriae Tres Auctore Aspasio Caramuelio Accessit Idiattibe de Prodigiosis Crucibus* (n.p.: n.p., 1664), p. 18. On mathematics as entertainment, see Menso Folkerts, "Unterhaltungsmathematik," in *Mass, Zahl und Gewicht. Mathematik als Schlüssel zu Weltverständnis und Weltbeherrschung*, exh. cat. (Wolfenbüttel: Herzog August Bibliothek, Acta Humaniora 1989), pp. 345–371.

73. Daniel Schwenter, *Deliciae Physico-Mathematicae. Mathemat.-und philosophische Erquickstunden. Darinnen sechshundert drey und sechsig schöne, liebliche und annehmliche Kunststücklein, auffgaben und fragen auf die Rechenkunst, Landtmessen, Perspectiv, Naturkündigung, und andern Wissenschaften genommen, begriffen seindt* (Nuremberg: Jeremia Dümler, 1651), pp. 1–3.

74. Ibid., pp. 4–6. Also see Filippo Bonanni [or Buonanni], *Ricreatione dell'occhio, e della mente nell'osservation delle chiocciole, proposta à curiosi delle opere della natura* (Rome: A Spese di Felice Cesarettu, 1661), pp. 5–7. On natural history as spectacle, see my "Voyeur or Observer? Enlightenment Thoughts on the Dilemma of Display," *Configurations*, 1 (Fall 1992), 93–126.

75. Georg Philipp Harsdörffer, *Delitiae Mathematicae et Physicae. Der mathematischen und philosophischen Erquickstunden*, 3 vols. (Nuremberg:

Gedruckt und Verlegt bey Jeremia Dümler, 1651), II, 2.

76. Ibid., p. 231. For a taxonomy of visual tricks, see Roger N. Shepherd, *Mind Sights: Original Visual Illusions, Ambiguities, and Other Anomalies, with a Commentary on the Play of Mind in Perception and Art* (New York: W. H. Freeman and Company, 1990), pp. 19–22, 122–123.

77. On Zahn, see Kenseth, *Age of the Marvelous*, p. 362.

78. Jacques Ozanam, *Récréations mathématiques et physiques, qui contiennent plusieurs problèmes d'arithmétique, de géométrie, d'optique, de gnomonique, de cosmographie, de méchanique, de pyrotechnie, & de physique. Avec un traité nouveau des horloges élémentaires,* 2 vols. (Paris: Chez Jean Jombert, 1694), I, ii–iv.

79. Daniel L. Collins, "Anamorphosis and the Eccentric Observer: Inverted Perspective and Construction of the Gaze," *Leonardo,* 25, no. 1 (1992), 77.

80. Gerd Gigerenzer, Zeno Swijtink, Theodore Porter, Lorraine Daston, John Beatty, and Lorenz Krüger, *The Empire of Chance: How Probability Changed Science and Everyday Life* (Cambridge: Cambridge University Press, 1989), pp. 11–13, 17.

81. Walter Pape, *Das literarische Kinderbuch. Studien zur Entstehung und Typologie* (Berlin: De Gruyter, 1981), pp. 55–56.

82. Chartier, "The Practical Impact of Writing," pp. 147–151.

83. Noël-Antoine La Pluche, *Le Spectacle de la nature, ou entretiens sur particularités de l'histoire naturelle, qui ont paru les plus propres à rendre les jeunes-gens curieux, & à former leur esprit,* 8 vols. (2d ed.; Paris: Chez la Veuve Estienne & Fils, 1749–1756), I, viii.

84. Rousseau, *Emile,* pp. 89–90.

85. Charles Rollin, *The Method of Teaching and Studying the Belles Lettres, or an Introduction to Languages, Poetry, Rhetoric, History, Moral Philosophy, Physicks, etc. . . . Designed More Particularly for Students in the Universities,* 4 vols., trans. from the French (2d ed.; Printed for A. Bettesworth, 1737), I, 6.

86. John Locke, *Some Thoughts Concerning Education* [1693], in *The Educational Writings,* ed. John William Adamson (Cambridge: Cambridge University Press, 1922), sect. 129, p. 102.

87. Ibid., sect. 149, p. 116.

88. Jean-Antoine Nollet, *Leçons de physique expérimentale,* 3 vols. (Paris: Chez les Frères Guérin, 1743), I, viii, xvii, xxv.

89. Joseph-Aignan Sigaud de La Fond, *Des merveilles de la nature,* 2 vols. (rev. ed.; Paris: Chez Delaplace, 1802), I, 88. On Sigaud, see Brockliss, *French Higher Education,* p. 4.

90. Henri Decremps, *Philosophical Amusements, or Easy and Instructive Recreations for Young People* (London: Printed for J. Johnson, 1790), preface.

91. Edmé Guyot, *Nouvelles récréations physiques et mathématiques, contenant ce qui a été imaginé de plus curieux dans ce genre, et ce qui se découvre journellement aux qu'elles on a joint leurs causes, leurs effets, la manière de les construire, & l'amusement qu'on en peut tirer pour étonner & surprendre agréablement,* 8 vols. (new ed.; Paris: Chez l'Auteur et Chez Gueffier, 1772), I, 56–64; VI, 119–122.

92. Jacques Ozanam, *Récréations mathématiques et physiques, qui contiennent les problêmes et les questions les plus remarquables et les plus propres à piquer la curiosité, tant des mathématiques que de la physique; & le tout traité d'une manière à la portée des lecteurs qui ont seulement quelques connoissances légères de ces sciences,* ed. Jean-Etienne Montucla, 4 vols. (Paris: Chez Cl. Ant. Jombert, Fils Aîné, 1778), I, v. 1.

93. Jacques Ozanam, *Récréations mathématiques et physiques, qui contiennent les problêmes et les questions les plus remarquables, et les plus propres à piquer la curiosité, tant des mathématiques que de la physique; le tout traité d'une manière à la portée des lecteurs qui ont seulement quelques connoissances légères de ces sciences,* ed. [Jean-Etienne Montucla], 4 vols. in 2 (Paris: Chez Firmin-Didot, 1790), I, iii–v. Also see Jacques Lacombe, *Dictionnaire encyclopédique des amusemens des sciences mathématiques et physiques; des procédés curieux des arts; des*

*tours récréatifs subtils de la magie blanche, & des découvertes ingenieuses & variées de l'industrie; avec l'explication de quatre vingt-six planches, & d'un nombre infini de figures qui y sont relatives* (Paris: Chez Panckoucke, Imprimeur-Libraire, 1792), pp. 31–82, "Aimant."

94. Martin Frobenius Ledermüller, *Mikroscopische Gemüths und Augen-Ergötzung; bestehend in ein hundert nach der Natur gezeichneten und mit Farben erleuchteten Kupfertafeln* (Bayreuth: Christian de Launoy, 1761), pp. 20, 48, 72, 119.

95. William Hooper, *Rational Recreations in Which the Principles of Numbers and Natural Philosophy Are Learned and Copiously Elucidated, by a Series of Easy, Entertaining, and Interesting Experiments. Among Which Are All Those Commonly Performed with the Cards,* 4 vols. (4th ed.; London: Printed for B. Law and Son, and G. G. and J. Robinson, 1794), I, v–vi.

96. For Chardin's genre pictures, see Anne Hollander, *Moving Pictures* (New York: Alfred A. Knopf, 1989), p. 216, and Philip Conisbee, *Chardin* (Lewisburg, Pa.: Bucknell University Press, 1986). Also see Rousseau, *Emile,* p. 51; and Locke, *Thoughts Concerning Education,* sect. 130, p. 103.

97. S. Roscoe, *John Newbery and His Successors (1740–1814): A Bibliography* (Cambridge: Five Owls Press, 1973), pp. 252–253. Also see J. H. Plumb, "The New World of Children," in *The Birth of a Consumer Society,* ed. Neil McKendrick, John Brewer, and J. H.

Plumb (London: Europa Publications Limited, 1982), n. 76, p. 301.

98. Tom Telescope, *The Newtonian System of Philosophy, Adapted to the Capacities of Young Gentlemen and Ladies, and Familiarized and Made Entertaining by Objects with Which They Are Intimately Acquainted: Being the Substance of Six Lectures Read to the Lilliputian Society by . . . And Collected and Methodized for the Benefit of the Youth of These Kingdoms, by Their Old Friend Mr. Newberry, in St. Paul's Church Yard; Who Has Also Added a Variety of Copper-Plate Cuts, to Illustrate and Confirm the Doctrines Advanced* (London: Printed for J. Newbery, 1761), pp. 1–2.

99. Ibid., p. 5. Also see Rousseau, *Emile,* p. 146.

100. Telescope, *Newtonian System,* pp. 100–101.

101. Nicholas Tucker, *The Child and the Book: A Psychological and Literary Exploration* (Cambridge: Cambridge University Press, 1981), p. 51. Also see Plumb, "New World of Children," pp. 291–295.

102. Sandra Harding, *Whose Science? Whose Knowledge? Thinking from Women's Lives* (Ithaca: Cornell University Press, 1991), pp. 5, 30; and, more generally, on the gendered character of natural knowledge, see Ludmilla Jordanova, *Sexual Visions: Images of Gender in Science and Medicine between the Eighteenth and Twentieth Centuries* (New York: Harvester Wheatsheaf, 1989).

103. Benjamin Martin, *The General Magazine of Arts and Sciences, Philosophical, Philological, Mathematical and Mechanical,* 2 vols. (London: Printed for W. Owen, 1755–1763), I, 1v–x. For this author's prolific books, tracts, and periodicals, see John R. Milburn, *Benjamin Martin: Author, Instrument-Maker, and "Country Showman"* (Leyden: Noordhoff International Publishers, 1976), pp. 65–83.

104. Hunt, *Newe Recreations,* "To the Reader."

105. Rousseau, *Emile,* p. 48.

106. On casuistry and the role of example in recreational guides, see Hunter, *Before Novels,* pp. 285–289.

107. Martin, *General Magazine,* I, 3.

108. Ibid., I, 310.

109. Benjamin Martin, *The Young Gentleman and Lady's Philosophy in a Continued Survey of the Works of Nature and Art; by Way of a Dialogue,* in *General Magazine,* II, 162.

110. Martin's educational treatises were also advertisements for his optical instruments; see *The Philosophical Grammar; Being a View of the Present State of Experimental Physiology, or Natural Philosophy. In Four Parts* (2d rev. ed.; London: Printed for John Noon, 1738), preface.

111. Plumb, "New World of Children," pp. 305–306.

112. Adam Walker, *A System of Familiar Philosophy: In Twelve Lectures, Being the Courses Usually Read by . . . Containing the Elements and the Practical Uses to Be Drawn from the Chemical Properties of Matter; the Principles and Application of Mechanics; of Hydrostatics; of Hydraulics; of Pneumatics; of Magnetism; of Electricity; of Optics; and of Astronomy. Including Every Material Modern Discovery and Improvement to the Present Time* (London: Printed for the Author, 1799), pp. ix–x, 3.

113. David Brewster, *A Treatise on the Kaleidoscope* (Edinburgh: Printed for Archibald Constable & Co.; and London: Longman, Hurst, Rees, Orme, & Brown, 1819), pp. 6–7, 113. For Brewster's later work on the stereoscope, see Robert Silverman, "Instrumental Representation and Perception in Modern Science: Imitating Human Function in the Nineteenth Century" (Ph.D. diss., University of Washington, 1992), chap. 2: "The Giant Eyes of Science: The Stereoscope and Photographic Depiction in the Nineteenth Century."

114. For romanticism's "inward specter show," see Terry Castle, "Phantasmagoria: Spectral Technology and the Metaphorics of Modern Reverie," *Critical Inquiry,* 15 (Autumn 1988), 49; and Terence Rees, *Theater Lighting in the Age of Gas* (London: The Society for Theater Research, 1978), p. 87. On the joining of the two traditions, see Laurence S. Lockridge, *The Ethics of Romanticism* (Cambridge: Cambridge University Press, 1989), pp. 21–25.

115. For a cursory overview of children's playbooks, see Gerard L'E. Turner, *Scientific Instruments and Experimental Philosophy, 1550–1850* (Aldershot: Variorum, 1990), pp. 377–398.

116. For the argument that the way teachers teach is often determined by their idea of the subject matter, see Susan Stodolsky, *The Subject Matters* (Chicago and London: University of Chicago Press, 1988).

117. John M. Moffatt, *The Book of Science, Second Series, Comprising Treatises on Chemistry, Metallurgy, Mineralogy, Crystallography, Geology, Oryctology, Meteorology* (London: Chapman and Hall, 1835), p. viii; and John Henry Pepper, *Scientific Amusements for Young People. Comprising Chemistry, Crystallization, Coloured Fires, Curious Experiments, Optics, Camera Obscura, Microscope, Kaleidoscope, Magic Lantern, Electricity, Galvanism, Magnetism, Aerostation, Arithmetic, etc.* (London: Routledge, Warne, and Routledge, 1861), p. 77.

*2*

## THE VISIBLE INVISIBLE

1. For an important study of mountebank quackery in the textual tradition, see Serge Soupel and Roger A. Hambridge, *Literature and Science and Medicine* (Los Angeles: Publications of the William Andrews Clark Memorial Library, 1982).

2. Wesley Trimpi, "Review of *The Cambridge History of Literary Criticism.* Vol. I: *Classical Criticism,*" *Ancient Philosophy,* 12 (1992), 508.

3. John Pococke, *The Machiavellian Moment* (Princeton: Princeton University Press, 1975), pp. 135–137, speaks of civic humanism's interest in elevating the citizen's mind through a high and generalizing art. Also see Elizabeth A. Bohls, "Disinterestedness and Denial of the Particular: Locke, Adam Smith, and the Subject of Aesthetics," in *Eighteenth-Century Aesthetics and the Reconstruction of Art,* ed. Paul Mattick, Jr. (Cambridge: Cambridge University Press, 1993), pp. 20–22.

4. See, for example, Stephen S. Hall, "How Technique Is Changing Science," *Science,* 257 (July 17, 1992), 344–349.

5. Robertson, *Mémoires récréatifs,* I, p. 178.

6. Ibid., pp. 149–150. On the French neoclassical sculptors' battle against Canovesque or Italian "idols" and "charlatanism of form," see Jacques de Caso, *David d'Angers: Sculptural Communication in the Age of Romanticism,* trans. Dorothy Johnson (Princeton: Princeton University Press, 1992), pp. 26, 38.

7. David Brewster, *Letters on Natural Magic, Addressed to Sir Walter Scott, Bart.* (London: John Murray, 1833), p. 2.

8. On the logic of the libertine, see Colette Cazenobe, *Le Système du libertinage de Crebillon à Laclos* (Oxford: Oxford University Press, 1991), p. 313. For the need to confront science as a variegated set of scientific practices inseparable from a wider cultural world, see Peter Galison, "Aufbau/Bauhaus: Logical Positivism and Architectural Modernism," *Critical Inquiry,* 16 (Summer 1990), 751.

9. Dena Goodman, "Public Sphere and Private Life: Toward a Synthesis of Current Historiographical Approaches to the Old Regime," *History and Theory,* 31, no. 1 (1992), 1. She is arguing against Lynn Hunt, "The Unstable Boundaries of the French Revolution," in *A History of Private Life,* vol. IV: *From the Fires of the Revolution to the Great War* (Cambridge and London: Belknap Press of Harvard University Press, 1990), pp. 13–14.

10. J. L. Heilbron, "The Measure of Enlightenment," in *The Quantifying Spirit in the Eighteenth Century,* ed. Tore Frängsmyr, J. L. Heilbron, and Robin E. Rider (Berkeley, Los Angeles, Oxford: University of California Press, 1990), pp. 208–210.

11. On the fair at Bezons and other *foires* and *fêtes de loges,* see Robert M. Isherwood, *Farce and Fantasy in the Old Regime* (Cambridge: Harvard University Press, 1989), pp. 23–24.

12. Craig Clunas, *Superfluous Things: Material Culture and Social Status in Early Modern China* (Urbana and Chicago: University of Illinois Press, 1991), p. 12.

13. Tomlinson, *Francisco Goya,* p. 220.

14. Steven Shapin and Simon Schaffer, *Leviathan and the Air-Pump: Hobbes, Boyle, and the Experimental Life* (Princeton: Princeton University Press, 1985), pp. 129–130.

15. Hollander, *Moving Pictures,* p. 15.

16. Mark Jones, ed., *Fake? The Art of Deception,* exh. cat. (London: British Museum Publications, 1990), p. 13.

17. Clunas, *Superfluous Things,* p. 114. For the antithetical northern tradition, embodied in Hendrick Goltzius's celebration of the artisanal power of his engraver's hand, see Walter S. Melion, "Memory and the Kinship of Writing and Picturing in the Early Seventeenth-Century Netherlands," *Word & Image,* 8 (January–March 1992), 65.

18. On Scot's thesis that the saints had been replaced by charmers in English popular magic before and after the Reformation, see Keith Thomas, *Religion and the Decline of Magic* (New York: Charles Scribner's Sons, 1971), p. 265.

19. Scot, *The Discovery of Witchcraft,* pp. 181, 185.

20. Samuel Rid, *The Art of Jugling or Legerdemaine. Wherein Is Deciphered All the Conveyances of Legerdemaine and Jugling, How They Are Effected, and Wherein They Chiefly Consist. Cautions to Beware of Cheating at Cards and Dice. The Detection of the Beggarly Art of Alcumistry. And the Foppery of Foolish Cousoning Charmes. All Tending to Mirth and Recreation, Especially for Those That Desire to Have Insight and Private Practice Thereof* (London: Printed by George Eld, 1614), n.p.

21. The Gypsy—like the sorcerer, witch, or quack—belongs to the worldwide network of conjure, i.e., those alternative cultural practices that challenge authority. See the fine study by David H. Brown, "Conjure/Doctors: An Exploration of a Black Discourse in America, Antebellum to 1940," *Folklore Forum,* 23, nos. 1/2 (1990), 5.

22. Felicity Nussbaum is studying how eighteenth-century Britain defined its national sexuality in opposition to the supposed libertinism found in warmer climates. See her "Prostitution, Body Parts, and Sexual Geography" (Clark Library Seminar on Imaging the Body, October 23–24, 1992).

23. William Hazlitt, "The Indian Jugglers," in *Essays,* ed. Frank Carr (London, Felling-on-Tyne, and New York: The Walter Scott Publishing Co., 1889), pp. 225, 227–228. Also see Stanley Jones, *Hazlitt: A Life* (Oxford: Clarendon Press, 1989), p. 102.

24. Hazlitt, "Indian Jugglers," p. 224.

25. Ibid., p. 217.

26. Hazlitt's wavering between popular, or coarse, and high, or sublimated, taste continues in current design debates. See Lev Marovich, "'Real' Wars: Esthetics and Professionalism in Computer Animation," *Design Issues,* 8 (Fall 1991), 19.

27. Klaus Herding, *Courbet: To Venture Independence,* trans. John William Gabriel (New Haven and London: Yale University Press, 1991), pp. 3, 32, 40. While Herding notes the theme of struggling in the paintings of Géricault, Daumier, and Courbet, he does not connect it to the *structure* of their pictures.

28. Rousseau, *Emile,* pp. 173–174. On Rousseau, see Marc Shell, *The Economy of Literature* (Baltimore and London: Johns Hopkins University Press, 1978), p. 117.

29. Marian Hannah Winter, *The Theatre of Marvels,* trans. Charles Melden (New York: Benjamin Blom, 1964), p. 18.

30. Lacombe, *Dictionnaire encylopédique des amusemens,* p. 270.

31. Magical "flick" book (France, c. 1750), Newberry Library, Chicago, accession no. 45-807. Also see Scot, *Discovery of Witchcraft,* pp. 194–195.

32. See, especially, the essays in Ken Hirschkop and David Shepherd, eds., *Bakhtin and Cultural Theory* (Manchester: Manchester University Press, 1989); and Gary Saul Morson and Caryl Emerson, *Mikhail Bakhtin: Crea-tion of a Prosaics* (Stanford: Stanford University Press, 1991).

33. On the eighteenth-century passion for gambling, see Nicole Castan, "The Public and the Private," in *History of Private Life,* III, 405–406.

34. For the metaphorical association of deception, paper currency, and combinable words as polyvalent signs dissociated from things, or the "money of fools," see Pape, "Heiliges Wort," p. 18. On the South Sea Bubble, see Pat Rogers, *Eighteenth-Century Encounters: Studies in Literature and Society in the Age of Walpole* (Sussex: The Harvester Press, and New Jersey: Barnes & Noble, 1985), pp. 151–163.

35. Henri-Gabriel Duchesne, *Dictionnaire de l'industrie, ou collection raisonnée des procédés utiles dans les sciences et dans les arts; contenant nombre de secrets curieux & interressants pour l'économie & les besoins de la vie; l'indication de différentes expériences à faire; la description de plusieurs jeux très singuliers & très amusants; les notices des découvertes & inventions nouvelles; les détails nécessaires pour se mettre à l'abri des fraudes & falsifications dans plusieurs objets de commerce & de fabriques,* 3 vols. (Paris: Chez Lacombe, Libraire, 1776), II, 254.

36. Franz Anton, Graf von Sporck, *Der sogenannte Sinn-Lehre und geistvolle Todtentanz* (Vienna: Gedruckt bey Johann Thomas Edlen von Trattnern, 1767), pl. 42. On Hogarth's renditions of the contradictory nature of life in a modern urban world, see Sean Shesgreen, *Hogarth and the Times-of-the-Day Tradition* (Ithaca and London: Cornell University Press, 1983), p. 133.

37. Lacombe, *Dictionnaire encyclopédique des amusemens,* pp. 1–2. Also see my "Art of Conjuring, or How the Romantic Virtuoso Learned from the Enlightened Charlatan," *Art Journal,* 52 (Summer 1993), 22–30.

38. Lacombe, *Dictionnaire encyclopédique des amusemens,* pp. 357–358.

39. Ibid., pp. 639–640.

40. [Henry Breslaw], *Breslaw's Last Legacy; or the Magical Companion: Containing All That Is Curious, Pleasing, Entertaining, and Comical. Selected from the Most Celebrated Masters of Deception; as Well with Slight of Hand, as with Mathematical Inventions; Wherein Is Displayed the Mode and Manner of Deceiving the Eye, as Practiced by Those Celebrated Masters of Mirthful Deceptions. Including the Various Exhibitions of Those Wonderful Artists Breslaw, Sieur Comus, Junas, etc. . . . The Whole to Form a Real Book of Knowledge in the Art of Conjuration, in Which Is Displayed the Way to Make the Air Balloon and Inflammable Air* (London: Printed for T. Moore, 1784), pp. vi, ix–xi, 26.

41. Ibid., pp. 44–47.

42. On Astley, see Martin Green and John Swan, *The Triumph of Pierrot: The Commedia dell'Arte and the Modern Imagination* (New York: Macmillan, 1986), pp. 5–6. On the rise of impresarios during the early eighteenth cen-

tury, see Pat Rogers, *Literature and Popular Culture* (Sussex: The Harvester Press, and New Jersey: Barnes & Noble, 1978), pp. 10–13.

43. Philip Astley, *Natural Magic: or Physical Amusements Revealed by . . . , Riding Master, Westminster Bridge, London; Great Part of Which Are Intended to Be Added to the Several Entertainments of the Above Place, for the Year 1785, Only* (London: Printed for the Author, 1785), p. 6.

44. Henri Decremps, *La magia bianca svelata; o, Spiegazione dei giucchi di mano sorprendenti,* trans. from the French (Messina: Grande Ospedale, 1793), pp. 3–5.

45. Ibid., p. 23. Also see Johann Samuel Halle, *Fortgesetzte Magie, oder Zauberkräfte der Natur, so auf den Nutzen und Belustigung angewandt worden,* 17 vols. (Berlin: Bey Joachim Pauli, 1788–1802), V, 323–325. Halle appears to be showing the "dead canary" trick, but describes "Der Kunstvogel," or the "artificial bird," who sings at the audience's bidding.

46. Henri Decremps, *The Conjuror Unmasked; or La Magie blanche dévoilée: Being a Clear and Full Explanation of All the Surprizing Performances Exhibited as Well in This Kingdom as on the Continent by the Most Eminent Professors of Slight of Hand* (London: Printed for and sold by T. Denton, 1785), p. 30.

47. Ibid., pp. 44, 46. Also see Ozanam, *Récréations mathématiques,* pp. 443, 438–439. On the importance of palingenesis to eighteenth-century alchemists, see Allen G. Debus, *The French Paracelsians: The Chemical Challenge to Medical and Scientific Tradition in Early Modern France* (Cambridge: Cambridge University Press, 1991), pp. 161–162.

48. David H. Solkin, "ReWrighting Shaftesbury: The *Air Pump* and the Limits of Commercial Humanism," in *Painting and the Politics of Culture: New Essays on British Art, 1700–1850,* ed. John Barrell (Oxford and New York: Oxford University Press, 1992), p. 176. Also see his *Painting for Money: The Visual Arts in the Public Sphere in Eighteenth-Century England* (New Haven and London: Yale University Press, 1993), pp. 214–246.

49. Henri Decremps, *Testament de Jérôme Sharp, professeur de physique amusante, où l'on trouve parmi plusieurs tours de subtilité qu'on peut exécuter sans aucune dépense, des préceptes & des exemples sur l'art de faire des chansons impromptu* (Paris: Chez Les Clapart, 1788), pp. 93–95. On the symbolic motions of the hand, see the excellent study by David McNeill, *Hand and Mind: What Gestures Reveal about Thought* (Chicago and London: University of Chicago Press, 1992), pp. 3, 36.

50. On the libertine sexual attitudes of Greuze's pictures, and of rococo painting in general, see Madelyn Gutworth, *The Twilight of the Goddesses:*

*Women and Representation in the French Revolutionary Era* (New Brunswick: Rutgers University Press, 1992), pp. 9, 39–40.

51. David Fraser, "Joseph Wright of Derby and the Lunar Society," in *Wright of Derby,* exh. cat. (New York: Metropolitan Museum of Art, 1990), p. 19.

52. On the role of science and its diffusion through British industrial society, see Albert Boime, *Art in the Age of Reason, 1750–1800* (Chicago and London: University of Chicago Press, 1987), pp. 238–239.

53. Halle, *Fortgesetzte Magie,* I, 2–3.

54. Ibid., VIII, 464–468.

55. Ibid., XI, 198, 243–246.

56. Jacques de Vaucanson, *Le Mécanisme du flûteur automate, présenté à messieurs de l'Académie Royale des Sciences par . . . , auteur de cette machine, avec la description d'un canard artificiel, mangeant, beuvant, digerant & se vuidant, épluchant ses aîles & ses plumes, imitant en diverses manières un canard vivant. Inventé par le même. Et aussi celle d'une autre figure, également merveilleuse, jouant du tambourin & de la flûte, suivant la rélation, qu'il en a donnée depuis son mémoire écrite* (Paris: Chez Jacques Guérin, Imprimeur-Libraire, 1738), pp. 3–4. On automata, also see Ozanam, *Récréations mathématiques,* pp. 102–109.

57. Henri-François Le Dran, *The Operations in Surgery of Mr. . . . , Senior Surgeon of the Hospital of La Charité, Consultant Surgeon to the Army, Member of the Academy of Surgery at Paris, and Fellow of the Royal Society in London. With Remarks, Plates of the Operations, and a Sett of Instruments by William Cheselden, Esq., Surgeon to the Royal Hospital at Chelsea, and Member of the Academy of Surgery at Paris* (London: Printed for C. Hitch and R. Dodsley, 1749), pp. 1, 7.

58. Jacques Lacombe, *Le Spectacle des beaux-arts, ou considérations touchant leur nature, leurs objets, leurs effets & leurs règles principales* (Paris: Chez Lottin, 1763), pp. 5, 10–11.

59. Hermione Almeida, *Romantic Medicine and John Keats* (New York and Oxford: Oxford University Press, 1991), p. 35.

60. On the development of "neutral facts," see Lorraine Daston, "Marvelous Facts and Miraculous Evidence in Early Modern Europe," *Critical Inquiry,* 18 (Autumn 1991), 93–124. On the enthusiasm for geometrical certainties and perspective machines within academies, see Martin Kemp, *The Science of Art* (New Haven and London: Yale University Press, 1990), pp. 165–166.

61. William Cheselden, *Osteographia, or the Anatomy of the Bones* (London: n.p., 1733), "To the Reader."

62. Joseph-Aignan Sigaud de La Fond, *Dictionnaire des merveilles de la nature,* 3 vols. (rev. ed.; Paris: Delaplace, An X), II, 332–333. Also see Salverte, *Des sciences occultes,* p. 257.

63. Jean-Claude Besuchet, *L'Anti-Charlatan, ou traitement raisonné de la maladie vénérienne d'après l'état actuel de la science; ouvrage outil aux practiciens, et mis à la portée des personnages étrangers à l'art de guérir* (Paris: Chez Gabon, Libraire, Latour, P. Mongie, Aîné, 1819), p. 63. On quacks, or *gens à secrets,* see Jan Goldstein, *Console and Classify: The French Psychiatric Profession in the Nineteenth Century* (Cambridge: Cambridge University Press, 1987), p. 72.

64. Hume, *Of Miracles,* p. 13. On forgeries as one of the effects of printing, see David Lowenthal, *The Past Is a Foreign Country* (Cambridge: Cambridge University Press, 1985), p. 290.

65. Gilles Deleuze, *The Logic of Sense,* trans. Mark Lester with Charles Stivale, ed. Constantine V. Boundas (New York: Columbia University Press, 1990), pp. 253–254.

66. Jacob Leupold, *Anamorphosis Mechanica Nova, oder Beischreibung dreyer neuen Machinen mit welchem sehr geschwinde und leichte auch von denjenigen so solcher Wissenschaft unerfahren, mancherley Bilder und Figuren können gezeichnet werden, dass sie ganz ungestalted und unkäntlich fallen* ([Leipzig]: Gedruckt bey Immanuel Tietzen, 1713). On Leupold's famous "deformed images," see Jacques-Mathurin Brisson, *Dictionnaire raisonné de physique par Mr. . . . , de l'Académie Royale des Sciences, maître de physique & d'histoire naturelle des Enfants de France, professeur royale de physique expérimentale au Collège Royale de Navarre & censeur royale,* 3 vols. (Paris: Hôtel de Thou, 1781–1789), I, 92.

67. Brewster, *Treatise on the Kaleidoscope,* pp. 135, 143–149. For Brewster and the romantic preoccupation with visual illusion, see Frederick Burwick, "Romantic Drama: From Optics to Illusion," in *Literature and Science,* ed. Stuart Peterfreund (Boston: Northeastern University Press, 1990), pp. 167–208.

68. Julia Kristeva, *Histoires d'amour* (Paris: Denoël, 1983), pp. 253–254.

69. Pietz, "Problem of the Fetish, II," p. 26.

70. Pomian, *Collectors and Curiosities,* p. 156.

71. Jacques Savery de Bruslons, *Dictionnaire universel de commerce: contenant tout ce qui concerne le commerce qui se fait dans les quatre parties du monde,* 4 vols. (2d ed.; Geneva: Chez Les Héritiers Cramer & Frères Philibert, 1742), II, 545, 547–548.

72. Ibid., I, 33; II, 1517.

73. On pickpocketing, see Henri Decremps, *Le Parisien à Londres, ou avis aux français qui vont en Angleterre, contenant le parallèle des deux plus grandes villes de l'Europe* (Amsterdam: Chez Maradan, 1789), pp. 99–105.

74. Montagu, *Roman Baroque Sculpture,* p. 52; and Kenseth, *Age of the Marvelous,* p. 42.

75. Savery de Bruslons, *Dictionnaire universel de commerce,* I, 830.

76. Kernan, *Printing Technology,* pp. 11–12. On the fetish, see Henri-Gabriel Duchesne and Pierre-Joseph Macquer, *Manuel de naturaliste. Ouvrage dédié à Mr. de Buffon, de l'Académie Française, etc., etc. Intendant du Jardin Royal des Plantes* (Paris: Chez G. Desprez, Imprimeur du Roi & du Clergé de France, 1771), p. 204.

77. On artists depicted with their students, see Paula Rea Radisch, *"Qui peut définir les femmes?* Vigée-Lebrun's Portraits of an Artist," *Eighteenth-Century Studies,* 25 (Summer 1992), 443. For contemporary issues of copying, see Joan L. Kirsch and Russell A. Kirsch, "Storing Art Images in Intelligent Computers," *Leonardo,* 23, no. 1 (1990), 99.

78. On the different practices with regard to the model in the academy and the private studio, see the excellent study by Candace Clements, "The Academy and the Other: *Les Grâces* and *Le Genre Galant,*" *Eighteenth-Century Studies,* 25 (Summer 1992), 474–481.

79. James Elkins, "From Original to Copy and Back Again," *British Journal of Aesthetics,* 33 (April 1993), 114.

80. Norman Bryson, *Tradition and Desire: From David to Delacroix* (Cam-

bridge: Cambridge University Press, 1984), p. 122, has suggested that for Ingres the meaning of the painting is always another painting. On the variants, also see Eldon N. Liere, "Ingres' *Raphael and the Fornarina:* Reverence and Testimony," *Arts Magazine,* 56 (December 1981), 108–115.

81. David Irwin, "Art versus Design: The Debate 1760–1860," *Journal of Design History,* 4, no. 4 (1991), 219–224.

82. Kim Sloan, "Drawing—A 'Polite Recreation' in Eighteenth-Century England," *Studies in Eighteenth-Century Culture,* ed. Harry C. Payne (Madison: University of Wisconsin Press, 1982), XI, 217–240.

83. For the connection between monkeys and *similitudo,* see Michael Camille, *Image on the Edge: The Margins of Medieval Art* (London: Reaktion Books, 1992), p. 12.

84. See the excellent historical review of this hostility to impersonation by Jonas Barish, *The Antitheatrical Prejudice* (Berkeley, Los Angeles, London: University of California Press, 1981), p. 161.

85. For the self-conscious ambivalence of the rococo, see William Park, *The Idea of the Rococo* (Newark: University of Delaware Press, 1993).

86. Richard Brilliant, *Portraiture* (Cambridge and London: Harvard University Press, 1992), p. 7.

87. Mary D. Sheriff, in her fine *Fragonard: Art and Eroticism* (Chicago and London: University of Chicago Press, 1990), pp. 153–184, demonstrates the willed look of these portraits. My point, however, is that they were parodies of common charlatan practices. See also Sheriff's "Invention, Resemblance, and Fragonard's *Portraits de Fantaisie,*" *Art Bulletin,* 69 (1987), 77–87.

88. For the sophisticated mixture of fact and fiction in eighteenth-century artists' biographies, see H. Perry Chapman, "Persona and Myth in Houbraken's Life of Jan Steen," *Art Bulletin,* 75 (March 1993), 140–141.

89. On the more general notion that biographies and portraits are always re-creations, see Richard Wendorf, *The Elements of Life: Biography and Portrait Painting in Stuart and Georgian England* (Oxford: Clarendon Press, 1990), p. 10.

90. Lacombe, *Dictionnaire encyclopédique des amusemens,* p. 388.

91. Hollander, *Moving Pictures,* pp. 232–233.

92. Reed Benhamou, "Imitation in the Decorative Arts of the Eighteenth Century," *Journal of Design History,* 4, no. 1 (1991), 3.

93. For recipes, see [Albrecht Ernst Friedrich Crailsheim], *Die hundert und eine Kunst. Oder vermischte Sammlung allerhand nützlich auch lustiger und scherzhafter Curiositäten* (n.p.: n.p.,

1761–1762), pp. 5, 7, 40. On the late eighteenth-century "quarrel of inventors," see Jean Chatelus, *Peindre à Paris au XVIIIe siècle* (Paris: Editions Jacqueline Jambon, 1991), pp. 72–75.

94. Lacombe, *Dictionnaire encyclopédique des amusemens,* pp. 389–390. On the nonspecialist's ability to produce quick copies cheaply, see Marcia Pointon, *Hanging the Head: Portraiture and Social Formation in Eighteenth-Century England* (New Haven and London: Yale University Press, 1993), p. 3.

95. Jens-Heiner Bauer, *Daniel Nikolaus Chodowiecki. Das druckgraphische Werk. Die Sammlung Wilhelm Burggraf zu Dohna-Schlobitten* (Hannover: Verlag Galerie J. H. Bauer, 1982), p. 33.

96. On the pantograph, pentograph, or *singe,* see Nicolas Bion, *Traité de la construction et des principaux usages des instruments mathématiques. Avec les figures nécessaires pour l'intelligence de ce traité* (2d rev. ed.; The Hague: Chez P. Husson, J. Swart, H. Scheurleer, J. van Deuren, R. Alberts, C. Le Vier, & F. Boucquet, 1723), p. 89.

97. Lacombe, *Dictionnaire encyclopédique des amusemens,* p. 590.

98. For the *je ne sais quoi,* see my "Beauty of the Invisible: Winckelmann and the Aesthetics of Imperceptibility," *Zeitschrift für Kunstgeschichte,* 41 (1980), 65–78. On Sièyes, a prominent figure in the events of 1789 and in the *Oath of the Tennis Court,* and, later, part of David's group in exile at Brussels, see Warren Roberts, *Jacques-*

*Louis David, Revolutionary Artist: Art, Politics, and the French Revolution* (Chapel Hill and London: University of North Carolina Press, 1989), pp. 51, 58, 191. Also see Antoine Schnapper, *David. Témoin de son temps* (Paris: Bibliothèque des Arts, 1980), pp. 286–288.

99. For the varieties of automata, see Constance Eileen King, *The Encyclopedia of Toys* (New York: Crown Publishers, 1978), pp. 91–104. On the self as art, see Domna Stanton, *The Aristocrat as Art: A Study of the Honnête Homme and the Dandy in Seventeenth- and Nineteenth-Century French Literature* (New York: Columbia University Press, 1980), pp. 108, 119–121.

100. On the trickster Ulysses, see J. Hillis Miller, *Illustration* (London: Reaktion Books, 1992), p. 129.

101. François de Salignac de La Mothe Fénélon, *Dialogues concerning Eloquence in General; and particularly That Kind Which is fit for the Pulpit: by the Late Archbishop of Cambrai. With His Letter to the French Academy, concerning Rhetoric, Poetry, History, and a Comparison between the Antients and the Moderns,* trans. William Stevenson (London: Printed by T. Wood, for J. Walthoe, Jr., 1722), pp. 12, 50, 59, 61.

102. Ibid., p. 14. Also see De Quincey, "Rhetoric," in *Selected Essays,* pp. 81–82.

103. Vincent, *Literacy and Popular Culture,* p. 157. For the sumptuary impli-

cations of the anti-gawdy diatribes, see [Jacques-Antoine Dulaure], *Pogonologia, or A Philosophical and Historical Essay on Beards,* trans. from the French (London: T. Cadell and Exeter, 1786), p. 82. On the Jesuits as the chief standard-bearers of garish pageants, see Barish, *Antitheatrical Prejudice,* pp. 163–164.

104. Hume, *Of Miracles,* p. 39.

105. For Rousseau on the real social consequences of spreading the arts and sciences, see Catherine Chevalley, "Should Science Provide an Image of the World?" in *Scientists and Their Responsibilities* (Canton, Mass.: Watson Publishing International, 1989), pp. 163–164. Also see Brewster, *Letters on Natural Magic,* p. 57.

106. Fénélon, *Dialogues concerning Eloquence,* p. 215.

107. Henri Decremps, *La Science sansculotisée. Premier essai sur les moyens de faciliter l'étude de l'astronomie, tant aux amateurs et aux gens de lettres, qu'aux marins de la République française et d'opérer une révolution dans l'enseignement* (Paris: Chez l'Auteur, An II), pp. 15–19.

108. Aimé-Henri Paullian, *Dictionnaire de physique dédié à Monseigneur le Duc de Berry,* 3 vols. (Avignon: Chez Louis Chambeau, 1761), II, 493–494. On Locke's French materialist disciples, see John Yolton, *Locke and French Materialism* (Oxford: Oxford University Press, 1991), p. 194.

109. On Cagliostro's complex political interaction with "irregular" Masons, Swedenborgians, and Illuminati, see Marsha Keith Schuchard, "The Secret Masonic History of Blake's Swedenborg Society," *Blake Studies,* 26 (Fall 1992), 40–48; and "William Blake and the Promiscuous Baboons," in *Consortium on Revolutionary Europe* (forthcoming).

110. On Machiavelli's espousal of "manly" Roman virtues, see Isaiah Berlin, *The Crooked Timber of Mankind: Chapters in the History of Ideas* (New York: Alfred A. Knopf, 1991), p. 9.

111. Jacques Derrida, *Margins of Philosophy* (Chicago: University of Chicago Press, 1982), p. 11.

112. [Roche de Marne], *Mémoires authentiques pour servir à l'histoire du comte de Caglyostro* (Paris: n.p., 1785), pp. 3–10.

113. Robertson, *Mémoires récréatifs,* I, 183, 191.

114. For a recent discussion of the affair, see Lynn Hunt, *The Family Romance of the French Revolution* (Los Angeles: University of California Press, 1992), pp. 103–106.

115. [Jeanne] Valois de La Motte, *Mémoires justificatifs de la comtesse de . . . . Ecrits par elle-même* ([Paris]: n.p., 1789), pp. 197–198.

116. Henri Decremps, *Philosophical Amusements, or Easy and Instructive Recreations for Young People* (London: Printed for J. Johnson, 1790), preface.

117. [Anon.], *Cagliostro démasqué à Varsovie. Ou rélation authentique des ses opérations alchimiques & magiques faites dans cette capitale en 1780 par un témoin oculaire* (n.p.: n.p., 1786), pp. 3–6.

118. [Thomas Andrew James], *Count Cagliostro: or, The Charlatan. A Tale of the Reign of Louis XVI,* 3 vols. (London: Edward Bull, Public Library, 1838), I, 107.

119. Ibid., I, 131.

120. Ibid., I, 134–135.

121. For Rousseau's polemic against Bernard Mandeville in the "great corruption debate," see Malcolm Jack, *Corruption and Progress: The Eighteenth-Century Debate* (New York: AMS Press, 1989), p. 183.

122. See Berlin's *Crooked Timber of Mankind,* p. 216, for a meditation on Kant's emphasis on human autonomy and the problems surrounding the moral worth of an act.

123. Rousseau, *Emile,* p. 41. Also see John Wiltshire, *Samuel Johnson in the Medical World: The Doctor and the Patient* (Cambridge: Cambridge University Press, 1991), p. 198.

124. See the excellent study of late medieval comic theater by Donald Maddox, *The Semiotics of Deceit: The Pathelin Era* (Lewisburg, Pa.: Bucknell University Press, 1984), p. 12.

125. Stuart E. Baker, "Georges Feydeau and the Aesthetics of Farce" (Ph.D. diss., City University of New York, 1976), pp. 18–20.

*3*

## LABORATORY GAMES

1. Leon Kass, *Toward a More Natural Science* (New York: The Free Press, 1985), pp. 213–217.

2. Ernest Gellner, *The Devil in Modern Philosophy,* ed. I. C. Jarvie and Joseph Agassi (London and Boston: Routledge & Kegan Paul, 1974), p. 5.

3. Timothy Murray, *Theatrical Legitimation: Allegories of Genius in Seventeenth-Century England and France* (New York: Oxford University Press, 1987), p. 194.

4. See my "Beauty of the Invisible: Winckelmann and the Aesthetics of Imperceptibility," pp. 65–78. For the different aesthetic implications of a theory of *generation* and one of *reproduction,* see my *Body Criticism,* chapter 3: "Conceiving."

5. The vitalism of alchemy is consonant with the physicality of craft. See Betty Jo Teeter Dobbs, *Alchemical Death and Resurrection: The Significance of Alchemy in the Age of Newton* (Washington, D.C.: Smithsonian Institution Libraries, 1990), pp. 3–4.

6. For recent discussions of the mid-eighteenth century as an age of art and craft and its reception as such in

the nineteenth century, see Deborah L. Silverman, *Art Nouveau in Fin-de-Siècle France: Politics, Psychology, and Style* (Los Angeles, Berkeley, and London: University of California Press, 1989), pp. 25–32; and Rae Beth Gordon, *Ornament, Fantasy, and Desire in Nineteenth-Century French Literature* (Princeton: Princeton University Press, 1992), pp. 7–15.

7. For the notion that invention is a mental or cognitive process, see W. Bernard Carlson and Michael E. Gorman, "A Cognitive Framework to Understand Technological Creativity: Bell, Edison, and the Telephone," in *Inventive Minds: Creativity in Technology,* ed. Robert J. Weber and David N. Perkins (New York and Oxford: Oxford University Press, 1992), pp. 48–79; and "Invention as Re-Representation: The Role of Sketches in the Development of the Telephone," Workshop on Cognitive Approaches to Visualization in Science and Technology (Princeton University, February 13, 1993).

8. Wilda Anderson, *Diderot's Dream* (Baltimore and London: Johns Hopkins University Press, 1990), p. 39.

9. Allan Franklin, *The Neglect of Experiment* (Cambridge: Cambridge University Press, 1986), pp. 165–168. Also see Lorraine Daston, *Classical Probability in the Enlightenment* (Princeton: Princeton University Press, 1988), pp. 92–93.

10. For the superior status accorded those mental disciplines aspiring to

spatial certitude, see the excellent study by Martin Kemp, *The Science of Art* (New Haven and London: Yale University Press, 1990), p. 39, and, on the importance of manuals of practical geometry, pp. 158–162.

11. Sebastien LeClerc, *Pratique de la géométrie, sur le papier et sur le terrain. Ou par une méthode nouvelle & singulière l'on peut avec facilité & en peu de temps se perfectionner en cette science* (Paris and Amsterdam: Chez Pierre Mortier, 1691), pp. 36–37.

12. Michael Lynch, "Laboratory Space and the Technological Complex: An Investigation of Topical Contextures," *Science in Context,* 4 (1991), 51–78.

13. This passage was singled out for approval by Joseph-Aignan Sigaud de La Fond, *Description et usage d'un cabinet de physique expérimentale,* 2 vols. (Paris: Chez P. Gueffier, 1775), I, xvi.

14. Pierre van Musschenbroek, *Cours de physique expérimentale et mathématique,* 3 vols., trans. Joseph-Aignan Sigaud de La Fond (Paris: Chez Bauche, 1769), I, xxxii. For the kinds of playful distortion the rigorous observer must avoid, see Christine Buci-Glucksmann, *La Folie du voir: De l'esthétique baroque* (Paris: Editions Galilée, 1986), p. 42

15. Salverte, *Des sciences occultes,* I, 139–140.

16. Louis Joblot, *Descriptions et usages de plusieurs nouveaux microscopes, tant simples que composez; avec des nouvelles*

observations faites sur une multitude innombrable d'insectes, & d'autres animaux de diverses espèces, qui naissent dans des liqueurs préparées, & dans celles qui ne le sont point* (Paris: Chez Jacques Collombat, 1718), pl. 6. On Joblot, see my "Picturing Ambiguity: Eighteenth-Century Microscopy and the Neither/Nor," in *Visions of Empire: Voyages, Botany, and Representations of Nature,* ed. David Miller (Cambridge: Cambridge University Press, 1994); and James Elkins, "On Visual Desperation and the Bodies of Protozoa," *Representations,* 40 (Fall 1992), 46.

17. Musschenbroek, *Cours de physique,* I, xvii–xviii, xx–xxi.

18. G. S. Rousseau, "Wicked Whiston and the English Wits," in *Enlightenment Borders: Pre- and Post-Modern Discourses: Medical, Scientific* (Manchester: Manchester University Press, 1991), pp. 330–331.

19. Bernard Forest de Bélidor, *Architectura Hydraulica, oder die Kunst, das Gewässer zu den verschiedentlichen Nothwendigkeiten des Menschlichen Lebens zu leiten, in die Höhe zu bringen, und vortheilhafftig anzuwenden,* 2 vols., trans. J. N. J. (Augsburg: Verlegts Johann Georg Mertz, Kunst-Händler, 1740–1748), I, *Vorrede,* and pp. 22, 51–52. On the need for close connections between engineers and the machine shop, see Eugene Ferguson, *Engineering and the Mind's Eye* (Cambridge and London: MIT Press, 1992), p. 58.

20. Also see Nollet, *Leçons de physique,* II, 202–203; III, 4.

21. Jean-Antoine Nollet, *L'Art d'expérience, ou avis aux amateurs de la physique, sur la choix, la construction et l'usage des instruments; sur la préparation et l'emploi des drogues qui servent aux expériences,* 3 vols. (2d ed., Paris: Chez P. E. G. Durand, Neveu, Libraire, 1770), I, xii–xxii.

22. Latour, *Science in Action,* pp. 64–66. Also see Andrew Pickering, ed., *Science as Practice and Culture* (Chicago and London: University of Chicago Press, 1992), pp. 5–8.

23. François-Saneré Cherubim d'Orléans, *La Dioptrique oculaire, ou la théorique, la positive, et la méchanique, de l'oculaire dioptrique en toutes ses espèces* (Paris: Chez Thomas Joly & Simon Benard, 1671), pp. 249–250.

24. Wilhelm Dorn, *Meil Bibliographie. Verzeichnis der vom dem Radierer Johann Wilhelm Meil illustrierten Bücher und Almanachen* (Berlin: Im Verlage von Gsellius, 1928), pp. 6–10.

25. Paullian, *Dictionnaire de physique,* I, 96.

26. Benjamin Carrard, *Essai qui a remporté le prix de la société hollandoise des sciences de Haarlem en 1770 sur cette question. Qu'est-ce qui est requis dans l'art d'observer; & jusques-où cet art contribue-t-il à perfectionner l'entendement?* (Amsterdam: Chez Marc-Michel Rey, 1777), 4–8.

27. Jean Senebier, *Essai sur l'art d'observer et de faire les expériences; . . . considérablement changée et augmentée,* 3 vols. (2d ed., Geneva: Chez J. J. Baschoud, Libraire, An X), I, 125.

28. Ibid., I, 25–26.

29. Frängsmyr et al., *The Quantifying Spirit,* p. 29. Also see Christian Wolff, *Allerhand nützliche Versuche, dadurch zu genauer Erkäntniss der Natur und Kunst der Weg gehäbnet wird, den Liebhabern Wahrheit mitgetheilt,* 3 vols. (Halle: Rengerischen Buchhandlung, 1721–1729), I, 4.

30. Senebier, *Essai sur l'art d'observer,* I, 29–30.

31. Antonio Pérez-Ramos, *Francis Bacon's Idea of Science and the Maker's Knowledge Tradition* (Oxford: Oxford University Press, 1988), p. 34.

32. For the intersection of science and etiquette in early modern Europe, see Mario Biagioli, *Galileo, Courtier: Science, Patronage, and Political Absolutism* (Chicago and London: University of Chicago Press, 1993).

33. Senebier, *Essai sur l'art d'observer,* I, 37–41.

34. Johann Georg Zimmermann, *Von der Erfahrung in der Arznenkunst* (Zurich: Bey Heidegger und Compagnie, 1763), p. 45.

35. Senebier, *Essai sur l'art d'observer,* III, 171–181, 211–216.

36. Gainsborough seems to allude to the *topos* of the slipped chiton. See Marina Warner, *Monuments and Maidens: The Allegory of the Female Form* (New York: Atheneum, 1985), p. 279.

37. Richard Wendorf has discussed Reynolds's fascination with apparent contradictions. See his *Elements of Life,* pp. 233–235.

38. Nicholas Penny, ed., *Reynolds,* exh. cat. (London: Royal Academy of Arts, 1986), pp. 224–225.

39. Oskar Bätschmann, *Einführung in die Kunstgeschichtliche Hermeneutik* (Darmstadt: Wissenschaftliche Buchgesellschaft, 1984), pp. 87–91.

40. Hemsterhuis, *Lettre sur la sculpture,* in *Oeuvres,* I, 13–14, 21–22.

41. Charles Bonnet, *Essai de psychologie; considérations sur les opérations de l'âme sur l'habitude et sur l'éducation, auxqu'elles on a ajouté des principes philosophiques sur la cause première et sur son effet* (London: n.p., 1755), pp. 238–239.

42. Stanton, *Aristocrat as Art,* pp. 150–155.

43. On the importance of melodrama in the late eighteenth century, see Pierre Frantz, "L'Espace dramatique de *La Brouette de vinaigrier à Coelina,*" *Revue des Sciences Humaines,* 162, no. 2 (1976), 171–182; and Jean-Paul Davoine, "L'Epithète mélodramatique," *Revue des Sciences Humaines,* 162, no. 2 (1976), 183–203.

44. See my *Body Criticism,* pp. 300–305, 410–413.

45. Shelley L. Frisch, "Poetics of the Uncanny: E. T. A. Hoffmann's 'Sandman,'" in R. Collins, *The Scope of the Fantastic* (Westport, Conn.: Greenwood Press, 1985), pp. 49–55.

46. For Bellmer, see Therese Lichtenstein, "Behind Closed Doors," *Artforum,* 29 (June 1991), 119–123; and Haim Finkelstein, "The Incarnation of Desire: Dalí and the Surrealist Object," *Res,* 23 (Spring 1993), 124–125, 136.

47. Veronica Hollinger, "Cybernetic Deconstructions: Cyberpunk and Postmodernism," *Mosaic,* 23 (Spring 1990), 32.

48. On Walter Benjamin's neurological understanding of modern experience, see Susan Buck-Morss, "Aesthetics and Anaesthetics: Walter Benjamin's Artwork Essay Reconsidered," *October,* 62 (Fall 1992), 16.

49. Nollet, *Leçons de physique,* I, xxxi.

50. Simon Schaffer, "Self-Evidence," *Critical Inquiry,* 18 (Winter 1992), 327–361; and "The Consuming Flame: Electrical Showmen and Tory Mysticism in the World of Goods," in *Consumption and the World of Goods,* ed. John Brewer and Roy Porter (London and New York: Routledge, 1993), pp. 489–526.

51. Baker, "Georges Feydeau and the Aesthetics of Farce," pp. 11–20. The recently identified distinction between nineteenth-century electricians such as William Sturgeon or E. M. Clarke and members of the professoriate such as Michael Faraday or Charles Wheatstone, I argue, originated in the eighteenth century. See Iwan Rhys Morus, "Currents from the Underworld: Electricity and the Technology of Display in Early Victorian England," *Isis,* 84 (March 1993), 52–54.

52. Rollin, *The Method of Teaching and Studying the Belles Lettres,* II, 256.

53. Also see my "Instructive Games: Apparatus and the Experimental Aesthetics of Imposture," in *Appearances: Essays in Culture, Perception, and the Arts,* ed. Walter Pape (forthcoming).

54. Henri Decremps, *Codicile de Jérôme Sharp, professeur de physique amusante; où l'on trouve, parmi plusieurs tours dont il n'est point parlé dans son testament, diverses récréations relatives aux sciences & beaux-arts* (4th ed., Paris and Liège: Chez E. J. Desoer, 1791), pp. 40–45.

55. Michael Fried, *Courbet's Realism* (Chicago and London: University of Chicago Press, 1990), p. 66. Also see his *Absorption and Theatricality: Painting and Beholder in the Age of Diderot* (Berkeley, Los Angeles, and London: University of California Press, 1980), pp. 4, 93, for the notion that the relation between painting as spectacle and the beholder was emerging as a problem in the middle of the eighteenth century.

56. For Kauffmann's designs used on furniture, one of which shows Abra making a wreath, see Malise Forbes Adam and Mary Mauchline, "*Ut Pictura Poesis:* Angelika Kauffmann's Literary Sources," *Apollo* (June 1992), pp. 347–348. Also see David Alexander, "Kauffmann and the Print Market in Eighteenth-Century England," in *Angelika Kauffmann: A Continental Artist in Georgian England,* ed. Wendy Wassyng Roworth (London: Reaktion Books, 1993), pp. 158–163.

57. Musschenbroek, *Cours de physique,* I, xxviii–xxix.

58. For the importance of science outside the walls, see Jan Golinski, *Science as Public Culture: Chemistry and Enlightenment in Britain, 1760–1820* (Cambridge: Cambridge University Press, 1992). For its antithesis, see Maurice Crosland, *Science under Control: The French Academy of Sciences, 1795–1914* (Cambridge: Cambridge University Press, 1992).

59. Brockliss, *French Higher Education,* pp. 189–192, 452–453.

60. This was also the case during and after the Revolution; see Mona Ozouf, *Festivals of the French Revolution,* trans. Alan Sheridan (Cambridge and London: Harvard University Press, 1988), pp. 21–26.

61. Joseph R. Roach, "Darwin's Passion: The Language of Expression on Nature's Stage," in *Discourse,* 13 (Fall-Winter 1990–1991), 40.

62. Duchesne, *Dictionnaire de l'industrie,* II, 282.

63. For Desaguliers's far-ranging business deals, see Larry Stewart, *The Rise of Public Science: Rhetoric, Technology, and Natural Philosophy in Newtonian Britain, 1660–1750* (Cambridge: Cambridge University Press, 1992), p. 392.

64. G. S. Rousseau, "Science Books and Their Readers in the Eighteenth Century," in *Books and Their Readers in Eighteenth-Century England,* ed. Isabel Rivers (Leicester: Leicester University Press, 1982), pp. 206–208.

65. On Savery, see Peter Mathias and John A. Davis, *Innovation and Technology in Europe: From the Eighteenth-Century to the Present Day* (Oxford: Blackwell, 1991), p. 8. Also see Stewart, *Rise of Public Science,* pp. 365–368.

66. Arnold Pacey, *The Culture of Technology* (Cambridge: MIT Press, 1983), p. 18.

67. John Theophilus Desaguliers, *A Course of Experimental Philosophy,* 2 vols. (2d rev. ed., London: Printed for W. Innys, T. Longman, T. Shewell, and C. Hitch, 1745), I, x–xi, 485–490.

68. Ibid., I, 265.

69. Hazlitt, *Essays,* p. 221.

70. Rid, *Art of Jugling,* preface.

71. Duchesne, *Notice historique sur della Porta,* p. 182; and Nicolas Bion, *The Construction and Principal Uses of Mathematical Instruments,* trans. Edmund Stone (2d ed., London: Printed for J. Richardson, 1758), p. 97.

72. De Quincey, "Rhetoric," in *Selected Essays,* pp. 116–117.

73. Rollin, *Method of Teaching and Studying the Belles Lettres,* II, 65–66. For the importance of Rollin in the secondary school curriculum, see Anthony Vidler, *Claude-Nicolas Ledoux: Architecture and Social Reform at the End of the Ancien Régime* (Cambridge and London: MIT Press, 1990), p. 7.

74. Rollin, *Method of Teaching and Studying the Belles Lettres,* II, 68–69. On enthusiasm from Locke to John Wesley, see Isabel Rivers, *Reason, Grace, and Sentiment: A Study of the Language of Religion and Ethics in England, 1660–1780,* vol. I (Cambridge: Cambridge University Press, 1991), pp. 34–35, 207, 239–240. In France, it takes the form of a sensualized empiricism. See Roger Mercier, "Sur le sensualisme de Rousseau. Sensation et sentiment dans la première partie des *Confessions,*" *Revue des Sciences Humaines,* 161, no. 1 (1976), 20.

75. Rollin, *Method of Teaching and Studying the Belles Lettres,* II, 81.

76. Sheriff, *Fragonard,* pp. 104–111. For the seemingly spontaneous style of Fragonard and Greuze, see Georges Vigarello, *Le Corps redressé. Histoire*

*d'un pouvoir pédagogique* (Paris: Jean-Pierre Delarge, Editeur, 1978), p. 52.

77. Carla Hesse, *Publishing and Cultural Politics in Revolutionary Paris, 1789–1810* (Berkeley, Los Angeles, Oxford: University of California Press, 1991), p. 170.

78. [François] Pelletier, *Mémoires du Sieur . . . , ingenieur-machiniste, sur les effets, propriétés & avantages d'une machine de son invention, concernant les armes à feu; & plusieurs autres qu'il a présentées depuis peu à l'Académie Royale des Sciences* (Paris: De l'Imprimerie de Bourgogne, 1790), p. 4.

79. Isherwood, *Farce and Fantasy,* pp. 55, 234.

80. Robert Marrone, *Body of Knowledge: An Introduction to Body/Mind Psychology* (New York: State University of New York Press, 1990), pp. xi–xiii.

81. Charles Rabiqueau, *Le Spectacle du feu élémentaire ou cours d'électricité expérimentale* (Paris: Chez Jombert, Knapen, Duchesne, 1753), pp. 1–2.

82. Ibid., pp. 3–4, 232, 123, 242.

83. See my *Body Criticism,* pp. 450–464; and Alan Gauld, *A History of Hypnosis* (Cambridge: Cambridge University Press, 1992), pp. 4–5.

84. Rabiqueau, *Le Spectacle du feu,* pp. 142–154, 186–187.

85. Charles Rabiqueau, *Prospectus du cabinet de Mr. Rabiqueau* (Paris: Chez L'Auteur, 1772), pp. 9–12.

86. Barbara Bücherl, "Franz Xaver Messerschmidt Charakterköpfe," in *Wunderblock. Eine Geschichte der moderne Seele,* ed. Jean Clair, Cathrin Pichler, and Wolfgang Pircher (Vienna: Löcker Verlag, 1989), pp. 55–56.

87. On the control of expression, see Jean-Jacques Courtine and Claudine Haroche, *Histoire du visage. Exprimer et faire ses emotions XVIe–début XIXe siècle* (Paris: Rivages/Histoire, 1988), pp. 17–19. Also see Albert Boime, "Géricault and Georget: Portraying Monomaniacs to Service the Alienist's Monomania," *Oxford Art Journal,* 14, no. 1 (1991), 80–84.

88. On the connection between the *roman noir* and Géricault's paintings of body fragments, see Nina Athanassoglou-Kallmyer, "Géricault's Severed Heads and Limbs: The Politics and Aesthetics of the Scaffold," *Art Bulletin,* 74 (December 1992), especially pp. 609–613.

89. Charles Rabiqueau, *Le Microscope moderne, pour débrouiller la nature par le filtre d'un nouvel alembic chymique ou l'on voit un nouveau méchanisme physique universel* (Paris: Chez l'Auteur et Demonville, 1781), pp. 2, 17.

90. Joseph Priestley, *The History and Present State of Electricity, with Original Experiments* (2d rev. ed., London: Printed for J. Dodsley, J. Johnson, J. Payne, and T. Cadell, 1769), p. 519.

Also see Golinski, *Science as Public Culture,* pp. 65–66; and Stuart Peterfreund, "Blake, Priestley, and the 'Gnostic Moment,'" in *Literature and Science,* ed. Stuart Peterfreund (Boston: Northeastern University Press, 1990), pp. 150–153.

91. Golinski, *Science as Public Culture,* pp. 96–97.

92. Walker, *System of Familiar Philosophy,* p. 357.

93. See Lissa Roberts, "Setting the Table: The Disciplinary Development of Eighteenth-Century Chemistry as Read through the Changing Structures of Its Tables," in *The Literary Structures of Scientific Argument,* ed. Peter Dear (Philadelphia: University of Pennsylvania Press, 1991), pp. 99–132; and Henry Guerlac, *Lavoisier—The Crucial Year: The Background and Origin of His First Experiments on Combustion in 1772* (Cambridge: Cambridge University Press, 1990).

94. Philip Stewart, *Engraven Desire: Eros, Image and Text in the French Eighteenth Century* (Durham and London: Duke University Press, 1992), pp. 67–72.

95. Joseph-Aignan Sigaud de La Fond, *Leçons sur l'économie animale,* 2 vols. (Paris: Chez Nicolas-Augustin Delalain, Libraire; and Dijon: Chez la Veuve Coignard, 1761), I, xiii–xv.

96. Paul Barolsky, *Michelangelo's Nose: A Myth and Its Makers* (University Park and London: Pennsylvania State

University Press, 1990), pp. 119–120, 140.

97. Joseph-Aignan Sigaud de La Fond, *Description et usage d'un cabinet de physique expérimentale,* 2 vols. (Paris: Chez P. Gueffier, 1775), I, xix–xx.

98. Maiorino, *Portrait of Eccentricity,* p. 27.

99. André Doyon and Lucien Liaigre, *Jacques Vaucanson. Méchanique du génie* (Paris: Presses Universitaires de France, 1966), p. x.

100. Pérez-Ramos, *Francis Bacon's Idea of Science,* pp. 48–50.

101. On the tensions between engineers and architects in France, see Antoine Picon, *French Architects and Engineers in the Age of Enlightenment,* trans. Martin Than (Cambridge: Cambridge University Press, 1988), pp. 107–108, 119.

102. Howard Gardner, *Art, Mind, and Brain: A Cognitive Approach to Creativity* (New York: Basic Books, 1982), pp. 227–229.

103. John Yolton, *Locke and French Materialism* (Oxford: Clarendon Press, 1991), pp. 35–36, 108–119, 115.

104. See, for example, Mark H. Johnson and John Morton, *Biology and Cognitive Development: The Case of Face Recognition* (Oxford: Blackwell, 1991), pp. 1–13.

105. Brisson, *Dictionnaire raisonné de physique,* I, 92, 196.

106. [Jacques de] Vaucanson, *Le Mécanisme du flûteur automate, présenté à messieurs de l'Académie Royale des Science par . . . , auteur de cette machine, avec la description d'un canard artificiel, mangeant, beauvant, digerant & se vuidant, épluchant ses aîles & ses plumes, imitant en diverses manières un canard vivant. Inventé par le même. Et aussi celle d'une autre figure, également merveilleuse, jouant du tambourin & de la flûte suivant la rélation donnée depuis son mémoire écrite* (Paris: Chez Jacques Guerin, Imprimeur-Libraire, 1738), p. 18.

107. Brockliss, *French Higher Education,* pp. 406–408.

108. Vaucanson, *Le Mécanisme du flûteur automate,* pp. 11–13, 19–20.

109. Doyon and Liaigre, *Jacques Vaucanson,* pp. 126–129.

110. David F. Channell, *The Vital Machine: A Study of Technology and Organic Life* (New York and Oxford: Oxford University Press, 1991), p. 43.

111. Linda Zionkowski, "Aesthetics, Copyright, and 'the Goods of the Mind,'" *British Journal for Eighteenth-Century Studies,* 15 (Autumn 1992), 163–174.

112. Senebier, *Essai sur l'art d'observer,* III, 222–223.

113. On the machinelike nature of twentieth-century sports performers,

see John M. Hoberman, "Creating the 'New Man': German and Soviet Sports Photographs between the Wars," in *This Sporting Life, 1878–1991,* exh. cat., ed. Ellen Dugan (Atlanta: High Museum of Art, 1992), n.p.

114. Rémy Saisselin, *The Enlightenment against the Baroque* (Berkeley, Los Angeles, and Oxford: University of California Press, 1992), pp. 3–6.

115. Dorothy and Roy Porter, *Patient's Progress: Doctors and Doctoring in Eighteenth-Century England* (Cambridge: Cambridge University Press, 1989), pp. 96–114.

116. Bruno Latour, *The Pasteurization of France,* trans. Alan Sheridan and John Law (Cambridge and London: Harvard University Press, 1988), p. 213.

117. Victor E. Neuburg, *Popular Education in Eighteenth-Century England* (London: The Woburn Press, 1971), p. 57. Also see Judy Egerton, *Wright of Derby,* exh. cat. (New York: Metropolitan Museum of Art, 1990), p. 10.

118. Rousseau, *Emile,* p. 179.

119. Gellner, *The Devil in Modern Philosophy,* pp. 27–31. On the privileging of language, also see Latour, *Science in Action,* p. 184.

120. Marcus Walsh, "The Superfoetation of Literature: Attitudes to the Printed Book in the Eighteenth Century," *British Journal for Eighteenth-*

*Century Studies,* 15 (Autumn 1992), 160–161.

121. Lissa Roberts, "A Word and the World: The Significance of Naming the Calorimeter," *Isis,* 82 (June 1991), 199–222.

122. Savery de Bruslons, *Dictionnaire universel de commerce,* I, 145; and Brisson, *Dictionnaire raisonné de physique,* I, 112.

123. Rousseau, *Emile,* p. 176. On macro-actors, see Michael Callon and Bruno Latour, "Unscrewing the Big Leviathan: How Actors Macro-Structure Reality and How Sociologists Help Them Do So," in *Advances in Social Theory and Methodology: Toward an Integration of Micro- and Macro-Sociologies,* ed. K. Knorr-Cetina and A. V. Cicourel (Boston, London, Henley: Routledge & Kegan Paul, 1981), pp. 285–286.

124. [Henry-François Le Dran], *Abrégé économique de l'anatomie du corps humain à la portée de toute personne qui veut se connoître & s'instruire en cette partie, ainsi que de tous ceux se destinent au grand art de guérir les malades* (Paris: Chez P. Fr. Didot, 1768), p. 111.

125. Isherwood, *Farce and Fantasy,* pp. 15–21.

126. For the northern valorization of *handelingh,* see the excellent study by Walter S. Melion, *Shaping the Netherlandish Canon: Karel van Mander's Schilder-Boek* (Chicago and London:

University of Chicago Press, 1991), pp. 58–60.

127. Jacob Leupold, *Theatrum Arithmetico-Geometricum, das ist: Schauplatz der Rechn-und-Mess-Kunst, darinnen enthalten dieser beyden Wissenschaften nöthige Grund-Regeln und Handgriffe so wohl, als auch die unterschiedene Instrumente und Maschinen, welche Theils in der Ausübung auf den Papier, Theils auch im Felde besonderen Vortheil geben können* (Leipzig: Bey Christophe Zunkel, 1727), pp. 17–18.

128. Georges Ifrah, *From One to Zero: A Universal History of Numbers* (New York: Viking, 1985), pp. 291–310.

129. Leupold, *Theatrum Arithmetico-Geometricum,* pp. 2–4. I am grateful to Ingrid Rowland for having allowed me to read her book-length manuscript, "Order and Abacus: Humanism and the Merchant Mind in Italy, 1500–1520." See especially the chapter on "Angelo Colocci, the Humanist Mechanic." Also see Basil S. Yamey, *Art & Accounting* (New Haven and London: Yale University Press, 1989), pp. 128–129.

130. On the importance of the arabesque to the romantics, see Werner Busch, *Die notwendige Arabeske. Wirklichkeitsaneigung und Stilisierung in der deutschen Kunst des 19 Jahrhunderts* (Berlin: Gebr. Mann Verlag, 1985), pp. 45–46. Also see Frances S. Connelly, "Poetic Monsters and Nature Hieroglyphics: The Precocious Primitivism of Philipp Otto Runge," *Art Journal,* 52 (Summer 1993), 31–39.

131. See my study of this "unconditional" semiotics: *Symbol and Myth: Humbert de Superville's Essay on Absolute Signs in Art* (Cranbury, N.J.: Associated University Presses, 1979), especially pp. 133–153.

132. On the importance of separating out ignorant journeymen from liberal artists, see Morris Eaves, *The Counter-Arts Conspiracy: Art in the Age of Blake* (Ithaca and London: Cornell University Press, 1992), p. 168.

133. On emblems as signs, see McNeill, *Hand and Mind,* pp. 37–38, 56–59, 61–72.

134. Carol Houlihan Flynn, "Running Out of Matter: The Body Exercised in Eighteenth-Century Fiction," in *The Languages of Psyche: Mind and Body in Enlightenment Thought,* ed. G. S. Rousseau (Berkeley, Los Angeles, Oxford: University of California Press, 1990), p. 161.

*4*

EXHIBITIONISM

1. See David Summers's discussion of the misinterpretation of Saussure in "Conditions or Conventions: or The Disanalogy of Art and Language," in *The Language of Art History,* ed. Salim Kemal and Ivan Gaskell (Cambridge: Cambridge University Press, 1992), p. 182.

2. Francesco Pellizi, "Multiple Cultures," *Res,* 22 (Autumn 1992), 5–9.

3. David Bianculli, *Teleliteracy: Taking Television Seriously* (New York: Continuum, 1992), pp. 23–24, and Herbert I. Schiller, *Culture, Inc.: The Corporate Takeover of Public Expression* (New York and Oxford: Oxford University Press, 1989), p. 134.

4. The connection between Hollywood and science in film technology has been explored by Gregg Mitman, "Cinematic Nature: Hollywood Technology, Popular Culture. and the Science of Animal Behavior, 1925–1940" (paper delivered at Workshop on Visualization, Princeton University, April 2, 1993).

5. See Carol Squiers, "Special Effects/ Violence at Benetton," *Artforum* (May 1992), 18–19; and Nancy Millman, "Controversy ad Infinitum," *Chicago Tribune* (December 3, 1992), section C, pp. 1–4.

6. See the essays in Ivan Karp and Steven D. Lavine, eds., *Exhibiting Cultures: The Poetics and Politics of Their Display* (Washington, D.C.: Smithsonian Institution Press, 1991).

7. See my "Voyeur or Observer? Enlightenment Thoughts on the Dilemmas of Display," *Configurations,* 1 (Fall 1992), 95–128. On voyeurism, see Laura Mulvey, "Visual Pleasure and Narrative Cinema," in *Film Theory and Criticism,* ed. Gerald Mast and Marshall Cohen (3d ed., London and New York: Oxford University Press, 1985), pp. 803–816.

8. Stewart, *Rise of Public Science,* pp. 107, 118.

9. Philip Fisher, *Making and Effacing Art: Modern American Art in a Culture of Museums* (New York: Oxford University Press, 1991), pp. 5–6.

10. Arthur Danto, *Beyond the Brillo Box: The Visual Arts in Post-Historical Perspective* (New York: Farrar, Straus, Giroux, 1992), p. 11.

11. Schiller, *Culture, Inc.,* p. 31.

12. For the nineteenth-century view of museums as places for amelioration, see Don Gifford, *The Farther Shore: A Natural History of Perception, 1798–1984* (New York: Atlantic Monthly Press, 1990), p. 128.

13. Latour, *Science in Action,* pp. 222–224.

14. Herbert Jaumann, "Was ist ein Polyhistor? Gehversuche auf einer verlassenen Terrain," *Studia Leibnitziana,* 22, no. 1 (1990), 76.

15. Gunnar Broberg, "The Broken Circle," in Frängsmyr et al., *The Quantifying Spirit,* pp. 49–50.

16. See P. N. Furbank, *Diderot: A Critical Biography* (New York: Knopf, 1991).

17. Oscar Kenshur, "'The Tumour of Their Own Hearts': Relativism, Aesthetics, and the Rhetoric of Demystification," in *Aesthetics and Ideology,* ed. George Levine (Rutgers, N.J.: Rut-gers University Press, 1993). Also see his "Encyclopedic Nominalism and the Theory of Interdisciplinarity" (paper delivered at the American Society for Eighteenth-Century Studies annual meeting, Pittsburgh, April 1991).

18. John Bender, "Rhetoricality: On the Modernist Return of Rhetoric," in *The Ends of Rhetoric: History, Theory, Practice,* ed. John Bender and David S. Wellbury (Stanford: Stanford University Press, 1990), pp. 21–25.

19. Anthony Grafton, "*Polyhistor* into *Philolog:* Notes on the Transformation of German Classical Scholarship, 1780–1850," *History of Universities,* III (London: Avebury, 1983), p. 165.

20. David Carroll, *Paraesthetics: Fou-cault/Lyotard/Derrida* (New York and London: Methuen, 1987), p. 24.

21. Marcia Fabianski, "Iconography of the Architecture of the Ideal *Musea* in the Fifteenth to the Eighteenth Centuries," *Journal of the History of Collections,* 2, no. 2 (1990), 120–125.

22. Eva Schutz, "Notes on the History of Collecting and of Museums in the Light of Selected Literature of the Sixteenth to the Eighteenth Centuries," *Journal of the History of Collections,* 2, no. 2 (1990), 208. On the formation of the Hapsburg collections at Ambras and Prague, see Horst Bredekamp, *Antikensehnsucht und Maschinenglauben. Die Geschichte der Kunstkammer und die Zukunft der Kunstgeschichte* (Berlin: Verlag Klaus Wagenbach, 1993), pp. 35–36.

23. Paula Findlen, "The Museum: Its Classical Etymology and Renaissance Genealogy," *Journal of the History of Collections,* 1, no. 1 (1989), 70–71.

24. Robertson, *Mémoires récréatifs,* I, 206.

25. Duchesne, *Dictionnaire de l'industrie,* I, 454.

26. Brockliss, *French Higher Education,* pp. 90–91.

27. Vigarello, *Le Corps redressé,* p. 106.

28. Hunter, *Before Novels,* pp. 201–204.

29. See Bertrand Goldgar, *The Curse of Party: Swift's Relations with Addison and Steele* (Lincoln: University of Nebraska Press, 1961), pp. 80–81; Brian McCrea, *Addison and Steele Are Dead: The English Department, Its Canon, and the Professionalization of Literary Criticism* (Newark: University of Delaware Press, 1990), pp. 25–26; and Joseph Addison, "The Coffee House," in *The Spectator, Selected Essays* (London and New York: Frederick Marne and Co., 1909), pp. 263–266.

30. Mary Hamer, *Signs of Cleopatra: History, Politics, Representation* (London and New York: Routledge, 1993), pp. 46–49. Hamer discusses this experiment from the vantage of the feminization and intensification of the gaze it occasioned.

31. Stephen Copley, "The Fine Arts in Eighteenth-Century Polite Culture,"

in Barrell, *Painting and the Politics of Culture*, pp. 23–24.

32. Mary Vidal, *Watteau's Painted Conversations: Art, Literature, and Talk in Seventeenth- and Eighteenth-Century France* (New Haven and London: Yale University Press, 1992), p. 95. The importance of conversation as a social phenomenon has been examined by Elise Goodman, *Rubens: The Garden of Love as a Conversatie à la Mode* (Amsterdam and Philadelphia: John Benjamin Publishing Co., 1992), pp. 32–33.

33. Ann Bermingham, "The Origins of Painting and the Ends of Art: Joseph Wright of Derby's *Corinthian Maid*," in Barrell, *Painting and the Politics of Culture*, p. 162.

34. Charles R. Bailey, "Attempts to Institute a 'System' of Secular Secondary Education in France, 1762–1789," in Leith, *Facets of Education*, p. 106.

35. Bernard Bovier de Fontenelle, *Entretiens sur la pluralité des mondes* (rev. ed., Amsterdam: Chez Pierre Mortier, 1701), p. 2. On the role of preselected models in the rhetorical exercises of the *collège*, see Anthony Vidler, *Claude-Nicolas Ledoux: Architecture and Social Reform at the End of the Ancien Régime* (Cambridge and London: MIT Press, 1990), pp. 8–9.

36. Fontenelle, *Entretiens*, pp. 3–5.

37. Ibid., pp. 8–10.

38. Rollin, *The Method of Teaching and Studying the Belles Lettres*, IV, 213–214, 224.

39. Catherine Monbeig Goguel, "Le Dessin encadré," *Revue de l'art*, 76 (1987), 25–28.

40. Priestley, *History and Present State of Electricity*, p. iii; and Walker, *System of Familiar Philosophy*, p. x.

41. Martin J. S. Rudwick, *Scenes from Deep Time: Early Pictorial Representations of the Prehistoric World* (Chicago and London: University of Chicago Press, 1992), pp. 14–16.

42. On Lister see Paolo Rossi, *The Dark Abyss of Time: The History of the Earth and the History of Nations* (Chicago and London: University of Chicago Press, 1984), pp. 3–6.

43. Martin Lister, *Historiae sive synopsis methodicae conchyliorum et tabularum anatomicarum* (Oxford: E. Typographeo Clarendomano, 1770), no pagination. For the concept of natural hieroglyphics, see my *Voyage into Substance: Art, Science, Nature, and the Illustrated Travel Account, 1760–1840* (Cambridge and London: MIT Press, 1984), pp. 305–319. For the aestheticization of specimens, note that Aldrovandi used pictures in place of *naturalia;* see Giuseppe Olmi, "Science-Honour-Metaphor: Italian Cabinets of the Sixteenth and Seventeenth Centuries," in *The Origins of Museums: The Cabinet of Curiosities in Sixteenth- and Seventeenth-Century Europe,* ed. Oliver Impey and

Arthur MacGregor (Oxford: Clarendon Press, 1985), pp. 5, 7.

44. Stafford, *Voyage into Substance,* pp. 69–72.

45. René-Antoine Ferchault de Réaumur, *Mémoires pour servir à l'histoire des insectes,* 6 vols. (Paris: De l'Imprimerie Royale, 1734–1755), I, 12.

46. Paullian, *Dictionnaire de physique,* III, 54–55.

47. Noël-Antoine La Pluche, *Le Spectacle de la nature, ou entretiens sur particularités de l'histoire naturelle, qui ont paru les plus propres à rendre les jeunes gens curieux, & à former leur esprit,* 9 vols. (2d ed., Paris: Chez la Veuve Estienne & Fils, 1749–1756), I, xxii.

48. Ibid., I, viii–ix.

49. Vidal, *Watteau's Painted Conversations,* p. 94.

50. La Pluche, *Le Spectacle de la nature,* I, 13–15.

51. Réaumur, *Mémoires des insectes,* I, 10.

52. Madeleine Pinault, *Le Peintre et l'histoire naturelle* (Paris: Flammarion, 1990), p. 30.

53. Réaumur, *Mémoires des insectes,* I, 8–9.

54. Ibid., I, 25–26, 517–518, 532–533.

55. Donna J. Harraway, *Simians, Cyborgs, and Women: The Reinvention of Nature* (New York: Routledge, 1990), pp. 188–191.

56. See, for example, William Ashworth, "Remarkable Humans and Singular Beasts," in Kenseth, *Age of the Marvelous,* pp. 131–137.

57. Schutz, "Notes on the History of Collecting," p. 208.

58. See the excellent study by Philip Sohm, *Pittoresco: Marco Boschini, His Critics, and Their Critiques of Painterly Brushwork in Seventeenth- and Eighteenth-Century Italy* (Cambridge: Cambridge University Press, 1991), pp. 52, 129, 186, discussing the evolution of the concept and practice over three centuries.

59. On the connection of apparently random brushstrokes to Marinist anamorphoses, see Sohm, *Pittoresco,* p. 140.

60. Inger Sigrin Thomsen, "The Rhetoric of Ruins: Sterne, Goethe, and the Novelist as Architect" (Ph.D. diss., University of Chicago, 1993). Also see my discussion of marbling in *Body Criticism,* pp. 199–210.

61. See my "'Illiterate Monuments': The Ruin as Dialect and Broken Classic," *The Age of Johnson,* 1 (1987), 1–30.

62. Jacqueline Lichtenstein, *La Couleur éloquente. Rhétorique et peinture à l'âge classique* (Paris: Flammarion, 1989), p. 143.

63. On Kleiner as a topographer working in the tradition of Leonard Knyff, see Pinault, *Le Peintre et l'histoire naturelle,* pp. 100, 119.

64. William J. Mitchell, *The Reconfigured Eye: Visual Truth in the Post-Photographic Era* (Cambridge and London: MIT Press, 1992), p. 7.

65. Arthur MacGregor, "'A Magazine of All Manner of Inventions': Museums in the Quest for 'Solomon's House' in Seventeenth-Century England," *Journal of the History of Collections,* 1 (1989), 207. Also see Steven Shapin, "The House of Experiment in Seventeenth-Century England," *Isis,* 79 (September 1988), 377, 381–388.

66. I have gleaned this from the documentation available in the archives at the Getty Center for the History of Art and the Humanities. On the Vatican's anti-Turkish activity, see Johns, *Papal Art and Cultural Politics,* pp. 8–11.

67. On the importance and pervasiveness of vitalism in the eighteenth century, see G. S. Rousseau, "The Perpetual Crisis of Modernism and Traditions of Enlightenment Vitalism: With a Note on Mikhail Bakhtin," in *The Crisis in Modernism: Bergson and the Vitalist Controversy,* ed. Frederick Burwick and Paul Douglass (Cambridge and London: Cambridge University Press, 1993), 76–97.

68. See Jaumann, "Was ist ein Polyhistor?," p. 77; and Conrad Wiedemann, "Polyhistors Glück und Ende. Von D. G. Morhof zum jungen Lessing," in *Festschrift Gottfried Weber,* ed. Heinz Otto Burger and Klaus von See (Bad Homburg, Berlin, and Zurich: Verlag Gehlen, 1967), p. 217.

69. See Sylvia Lavin, *Quatremère de Quincy and the Invention of a Modern Language of Architecture* (Cambridge and London: MIT Press, 1992), pp. 126–127, 137, 143; and my *Symbol and Myth,* pp. 95–115.

70. Donald P. Rutherford, "Phenomenalism and the Reality of Body in Leibniz's Later Philosophy," *Studia Leibnitziana,* 22, no. 1 (1990), 17–19.

71. Paolo Rossi, "The Twisted Roots of Leibniz's 'Characteristic,'" in *The Leibniz Renaissance: International Workshop,* ed. Paolo Rossi and Walter Bernardi (Florence: Leo S. Olschki, 1989), pp. 274–278.

72. Stafford, "'Illiterate Monuments,'" p. 29.

73. Silverman, *Art Nouveau,* pp. 24–25.

74. Iain Pears, *The Discovery of Painting: The Growth of Interest in the Arts in England, 1680–1768* (New Haven and London: Yale University Press, 1988), pp. 114–123.

75. Sigaud de La Fond, *Des merveilles de la nature,* I, 297, 375. For André Breton's eclectic collections, see Leslie

Camhi, "Extended Boundaries," *Art in America* (February 1992), pp. 39–40.

76. Sigaud de La Fond, *Des merveilles de la nature,* I, iv.

77. On Grew, see Kenseth, *Age of the Marvelous,* pp. 97, 245.

78. Pomain, *Collectors and Curiosities,* pp. 54–55. Also see Lorraine Daston, "Marvelous Facts and Miraculous Evidence in Early Modern Europe," *Critical Inquiry,* 18 (Autumn 1991), 95.

79. Nehemiah Grew, *Musaeum Regalis Societatis, or a Catalogue & Description of the Natural and Artificial Rarities Belonging to the Royal Society and Preserved at Gresham College. Whereunto Is Subjoyned the Comparative Anatomy of Stomachs and Guts. By the Same Author* (London: Printed by W. Rawlins, 1681), preface.

80. Ibid., section VI.

81. Naomi Schor, *Reading in Detail: Aesthetics and the Feminine* (New York and London: Methuen, 1987), p. 5.

82. Friedrich Schiller, *On the Aesthetic Education of Man: In a Series of Letters,* ed. and trans. Elizabeth M. Wilkinson and L. A. Willoughby (Oxford: Clarendon Press, 1967), pp. 32–33.

83. For the evolution of the *cabinet d'histoire naturelle,* see Yves Laissus, "Le Jardin du roi," in *Enseignement et diffusion des sciences en France au XVIIIe siècle,* ed. René Taton (Paris: Hermann, 1964), pp. 296–297.

84. Ibid., pp. 287–295.

85. Pinault, *Le Peintre et l'histoire naturelle,* p. 40.

86. Elizabeth Liebman, "Recognizing Nature: The Role of Images in Eighteenth-Century Natural History" (B.A. Honors Paper, University of Chicago, Spring 1992). She is preparing an M.A. thesis on de Sève's unpublished drawings for the *Histoire naturelle.*

87. Pomian, *Collectors and Curiosities,* p. 125. Also see Walter E. Houghton, "The English Virtuoso in the Seventeenth Century," *Journal of the History of Ideas,* 3 (1942), 190–219.

88. On this question, see Csikszentmihalyi and Rochberg-Halton, *Meaning of Things,* pp. 188–190.

89. Louis-Jean-Marie Daubenton, "Cabinet d'histoire naturelle," in *Encyclopédie, ou Dictionnaire raisonné des sciences, des arts, et des métiers, par une société de gens de lettres mis en ordre et publié par M. Diderot; et quant à la partie mathématique, par M. D'Alembert,* 17 vols. (Paris: Le Breton, Briasson, David l'Aîné et Durand, 1759–1772), II, 488.

90. Georges-Louis Leclerc, Comte de Buffon and Louis-Jean-Marie Daubenton, *Histoire naturelle, générale et particulière, avec la description du cabinet du roy,* 15 vols. (Paris: De l'Imprimerie Royale, 1749–1767), III, 191–193.

91. Daubenton, "Cabinet d'histoire naturelle," *Encyclopédie,* II, 492.

92. Ibid., II, 490.

93. Ibid., II, 490–491.

94. Duchesne offered a lengthy commentary on the curatorial problems of conservation in the *Dictionnaire de l'industrie,* I, 265–272.

95. Schiller, *On the Aesthetic Education of Man,* p. 19.

96. Peter Vergo, "The Reticent Object," in Vergo, *New Museology,* p. 48. Also see J. M. Bernstein, *The Fate of Art: Aesthetic Alienation from Kant to Derrida and Adorno* (University Park, Pa.: Pennsylvania State University Press, 1992), p. 209.

97. John Lesch, "Systematics and the Geometrical Spirit," in Frängsmyr et al., *The Quantifying Spirit,* p. 111.

98. Joseph-Adrien Le Large de Lignac, *Lettres à un amériquain sur l'histoire naturelle de Mr. de Buffon; et sur les observations microscopiques de Mr. Needham* (Hamburg: n.p., 1751), pp. 42–49.

99. On the virtuoso, see Pears, *Discovery of Painting,* pp. 182, 201.

100. Lignac, *Lettres,* pp. 10, 15, 28–30.

101. Henri-Gabriel Duchesne and Pierre-Joseph Macquer, *Manuel de naturaliste. Ouvrage dédié à M. de Buffon, de l'Académie Française, etc., etc., Intendant du Jardin Royal des Plantes* (Paris: Chez G. Desprez, Imprimeur du Roi & du Clergé de France, 1771), p. v.

102. Wolf Lepenies, *Das Ende der Naturgeschichte. Wandel Kulturwelter Selbstverständlichkeiten in den Wissenschaften des 18. und 19. Jahrhunderts* (Munich and Vienna: Carl Hauser Verlag, 1976), pp. 30–31.

103. Joseph Rykwert, "Organic and Mechanical," *Res,* 22 (Autumn 1992), 11.

104. Lignac, *Lettres,* pp. 73–75.

105. Gifford, *Farther Shore,* p. 10. Also see Hubert Damisch, *Le Jugement de Pâris. Iconologie analytique 1* (Paris: Flammarion, 1992), p. 156.

106. Fisher, *Making and Effacing Art,* p. 24. On the rise of the systematic documentation of works of art, see W. McAllister Johnson, "From Verrue to Vence: Systematic Engraving of Private Painting Collections in France to 1760," *Gazette des Beaux-Arts,* 117 (February 1991), 78–84.

107. Charles Saumarez Smith, "Museums, Artifacts, and Meanings," in Vergo, *New Museology,* pp. 7–8. For comparison also see Martin Welch, "The Foundation of the Ashmolean Museum," in *Tradescant's Rarities: Essays on the Foundation of the Ashmolean Museum, 1683, with a Catalogue of the Surviving Early Collections,* ed. Arthur MacGregor (Oxford: Clarendon Press, 1983), pp. 40–58.

108. John and Andrew van Rymsdyck, *Museum Britannicum, Being an Exhibition of a Great Variety of Antiquities and Natural Curiosities, Belong-* ing to That Noble and Magnificent Cabinet, the British Museum, Illustrated with Curious Prints, Engraved after the Original Designs, from Nature, Other Objects; and with Distinct Explanations of Each Figure (London: Printed by I. Moore for the Authors, 1778), pp. iii–iv.

109. Jon P. Klancher, *The Making of English Reading Audiences, 1790–1832* (Madison: University of Wisconsin Press, 1987), p. 3. Also see John and Andrew van Rymsdyk, *Museum Britannicum; or a Display in Thirty-Two Plates, in Antiquities and Natural Curiosities, in That Noble Cabinet, the British Museum,* ed. P. Boyle (2d rev. ed., London: Printed for the Editor by J. Moore, 1791).

110. Rymsdyck, *Museum Britannicum* (1778), p. iv.

111. Ibid., pp. vii–viii.

112. Ibid., pp. 56–57.

113. Ibid., pp. 18–19, 51.

114. Ibid., p. xi. Also see Albert Heinekamp, "Das Gluck als höchstes Gut in Leibniz' Philosophie," in Rossi and Bernardi, *The Leibniz Renaissance,* p. 109.

115. Rymsdyck, *Museum Britannicum* (1778), pp. 10–13.

116. Latour, *Science in Action,* p. 219.

117. Sydney Pokorny, "Media Kids," *Artforum,* 31 (April 1993), 14.

118. Barbara Kirshenblatt-Gimblett, "Objects of Ethnography," in *The Poetics and Politics of Museum Display,* ed. Ivan Karp and Steven D. Lavine (Washington, D.C.: Smithsonian Institution, 1990), pp. 430–434. Also see Rémy Saisselin, *The Bourgeois and the Bibelot* (New Brunswick, N.J.: Rutgers University Press, 1984), pp. 68–71; and Fisher, *Making and Effacing Art,* p. 146.

## CONCLUSION

### THE TRANSIT OF INFORMATION

1. Dick Hebdige, *Hiding in the Light: On Images and Things* (London and New York: Routledge, 1988), p. 85.

2. Martin Jay, "The Rise of Hermeneutics and the Crisis of Ocularcentrism," in *Force Fields: Between Intellectual History and Cultural Critique* (New York and London: Routledge, 1993), pp. 100–101.

3. Postman, *Amusing Ourselves to Death,* pp. 7, 24, 28.

4. Saisselin, *The Enlightenment against the Baroque,* pp. 3–6.

5. Daston, "Marvelous Facts and Miraculous Evidence," pp. 117–118.

6. On Descartes, see Ernest Gellner, *Reason and Culture: The Historic Role of Rationality and Rationalism* (Oxford: Blackwell, 1992), p. 13.

7. For the corruption that Montaigne, Diderot, Voltaire, Rousseau, Mandeville, and Gibbon saw in social institutions, see Jack, *Corruption and Progress,* pp. 197–199.

8. For the radical difference between readers and listeners, see Kernan, *Printing Technology,* p. 220.

9. Klancher, *Making of English Reading Audiences,* p. 98.

10. De Quincey, "Style," in *Selected Essays,* p. 149.

11. G. S. Rousseau, "Science Books and Their Readers in the Eighteenth Century," in *Books and Their Readers in Eighteenth-Century England,* ed. Isabel Rivers (Leicester: Leicester University Press, 1982), pp. 207–208; and "Wicked Whiston and the English Wits," in *Enlightenment Borders: Pre- and Post-Modern Discourses* (Manchester: Manchester University Press, 1991), p. 330. For a similar growing distinction in France between applied and pure science books, see Jean DHombres, "Books: Reshaping Science," in *Revolution in Print: The Press in France 1775–1800,* ed. Robert Darnton and Daniel Roche (New York: New York Public Library, 1989), pp. 178–179.

12. On isolation as a social system, originating in the late eighteenth century with Jeremy Bentham and Adam Smith, see John Bender, *Imagining the Penitentiary: Fiction and the Architecture of Mind in Eighteenth-Century England*

(Chicago and London: University of Chicago Press, 1987), pp. 207–208.

13. Justine Cassell and David McNeill, "Gesture and the Poetics of Prose," *Poetics Today,* 12 (Fall 1991), 375–404.

14. Sander L. Gilman, *The Jew's Body* (New York: Routledge, 1991), pp. 128–149; and "Mark Twain and the Diseases of the Jews," *American Literature,* 65 (March 1993), 106, 110–112. Also see Hal Foster, "That Obscure Subject of Desire," in *Interim,* exh. cat. (New York: New Museum of Contemporary Art, 1990), p. 54.

15. Howard Gardner, *The Unschooled Mind: How Children Think and How Schools Should Teach* (New York: Basic Books, 1991), pp. 3–5.

16. Howard Gardner, *Art Education and Human Development,* Occasional Paper 3 (Los Angeles: The J. Paul Getty Trust, 1990), p. 49.

17. Ibid., pp. 9–13. Also see his *Frames of Mind,* pp. 8–9; and *Art, Mind, and Brain,* p. 240.

18. Ronald Kotulak, "Mental Workouts Pump Up Brain Power," *Chicago Tribune* (April 12, 1993), pp. 1, 12.

19. Mitchell, *Reconfigured Eye,* p. 13.

20. Claire Richter Sherman, "Beyond the Photo Archive: Imaging the History of Psychology," *Visual Resources,* 9 (1993), 39–41.

21. See my "Present Image, Past Text, Post Body: Educating the Late Modern Citizen," *Semiotica,* 91, nos. 3–4 (1992), guest editorial, 195–198.

22. Homi K. Bhabha, "Freedom's Basis in the Indeterminate," *October,* 61 (Summer 1992), pp. 48–50.

23. Susan S. Stodolsky, *The Subject Matters: Classroom Activity in Math and Social Sciences* (Chicago and London: University of Chicago Press, 1988), pp. 3, 135.

24. On the desire to approach the picture in order to read it, see Alan Jones, "Literature: The Lust for Words," *Arts Magazine,* 66 (April 1992), 27–28; and Michèle Cone, "Unpainting: Review of Thierry de Duve, *Pictorial Nominalism: On Marcel Duchamp's Passage from Painting to the Ready-Made,*" *Arts Magazine,* 66 (February 1992), 31.

25. See my essay "The Eighteenth Century at the End of Modernity: Towards the Re-Enlightenment," *Studies in Eighteenth-Century Culture,* 25 (1994), forthcoming.

26. Tony Kelly, "Technology, Interest Are Forcing Newspapers to Be Image-Conscious," *Chicago Journalist,* 3 (January–February 1993), 1, 4–5.

27. Carol Vogel, "Pop-Up Art Tutorial Has Projects for Everyone," *New York Times* (January 22, 1993), section B, 1, 8.

28. Rivers, *Reason, Grace, and Sentiment,* I, 34, 232–240.

29. Uli Schmetzer, "Reprise of Magic Cults a Stage Sizzler in China," *Chicago Tribune* (December 21, 1991), sect. 1, 6.

30. David Lomas, "Modernism *versus* Postmodernism: Kirk Varnedoe's 1992 Slade Lectures," *Apollo,* 137 (February 1993), 73–74.

31. On theft, see Pat Rogers, *The Augustan Vision* (London: Weidenfeld and Nicolson, 1974), pp. 99–102; and Peter Linebaugh, *The London Hanged: Crime and Civil Society in the Eighteenth Century* (Cambridge: Cambridge University Press, 1991).

32. Gellner, *The Devil in Modern Philosophy,* p. 6.

33. Fisher, *Making and Effacing Art,* pp. 212–213.

34. Carol M. Armstrong, "Edgar Degas and Representation of the Female Body," in *The Female Body in Western Culture: Contemporary Perspectives,* ed. Susan Rubin Suleiman (Cambridge: Cambridge University Press, 1986), p. 225.

35. Degas's strategy has less to do with the "gentrification of work" than with providing physiologies of doing. For the former interpretation, see Carol Armstrong, *Odd Man Out: Readings of the Work and Reputation of Edgar Degas* (Chicago and London: University of Chicago Press, 1991), p. 34.

36. Enzio Manzini, "Prometheus of the Everyday: The Ecology of the Artificial and the Designer's Responsibility," *Design Issues,* 9 (Fall 1992), 5–20.

37. On the eighteenth-century victory of masculine over feminine styles of doing science, see Londa Schiebinger, *The Mind Has No Sex? Women in the Origins of Modern Science* (Cambridge: Harvard University Press, 1989), pp. 121–125, 146.

38. Ada Louise Huxtable, "Inventing American Reality," *New York Review of Books* (December 3, 1992), pp. 24–29.

39. Mona Ozouf, *Festivals and the French Revolution,* trans. Alan Sheridan (Cambridge: Harvard University Press, 1988), pp. 206–207.

40. Carol Duncan, "Museums and the Ritual of Citizenship," in Karp and Lavine, *Poetics and Politics* of *Museum Display,* p. 93.

41. Silverman, *Art Nouveau,* p. 76.

42. Lionel Grossman, *Between History and Literature* (Cambridge and London: Harvard University Press, 1990), p. 193.

43. Schor, *Reading in Detail,* pp. 16–20.

44. Igor Kopytoff, "The Cultural Biography of Things: Commoditization as Process," in *The Social Life of Things: Commodities in Cultural Perspective,* ed. Arjun Appadurai (Cambridge:

Cambridge University Press, 1986), pp. 21, 70–74.

45. Gerald N. Izenberg, *Impossible Individuality: Romanticism, Revolution, and the Origins of Modern Selfhood, 1787–1802* (Princeton: Princeton University Press, 1992), p. 3.

46. On the negative poetics of the Jena Circle, see Philippe Lacoue-Labarthe and Jean-Luc Nancy, *The Literary Absolute: The Theory of Literature in German Romanticism* (Albany: State University of New York Press, 1988), p. 8.

47. Jerome J. McGann, *The Romantic Ideology: A Critical Investigation* (Chicago and London: University of Chicago Press, 1983), p. 71.

48. Schiller, *On the Aesthetic Education of Man,* p. 9.

49. For a Marxist critique of the modern self as aesthetic artifact, see Terry Eagleton, *The Ideology of the Aesthetic* (Oxford: Basil Blackwell, 1990), p. 3.

50. Werner Busch, *Die notwendige Arabeske. Wirklichkeitsaneigung und Stilisierung in der deutschen Kunst des 19. Jahrhunderts* (Berlin: Gebr. Mann Verlag, 1985), p. 46.

51. Schiller, *On the Aesthetic Education of Man,* p. 97.

52. Jeffrey Barnouw, "The Beginnings of 'Aesthetics' and the Leibnizian Conception of Sensation," in Mattick, *Eighteenth-Century Aesthetics,* pp. 94–95.

53. Rae Beth Gordon, *Ornament, Fantasy, and Desire in Nineteenth-Century French Literature* (Princeton: Princeton University Press, 1992), p. 15.

54. Joseph Leo Koerner, "Borrowed Sight: The Halted Traveller in Caspar David Friedrich and William Wordsworth," *Word & Image,* 1 (April-June 1985), 151.

55. For Friedrich as a conjuror, not a creator, of situations, see Hollander, *Moving Pictures,* pp. 294–298.

56. Shaffer, *'Kubla Khan' and the Fall of Jerusalem,* p. 189.

57. It seems to me that there is a basic connection between romantic pictorial strategies and the electronic environment. See Roy Ascott and Carl Eugene Loeffler, "Connectivity: Art and Interactive Telecommunications," *Leonardo,* 24, no. 2 (1991), 115–117.

58. Roger Mercier, "Sur le sensualisme de Rousseau. Sensation et sentiment dans la première partie des *Confessions,*" in *Revue des Sciences Humaines,* 161, no. 1 (1976), 20.

59. See Izenberg, *Impossible Individuality,* pp. 55–56; Lockridge, *Ethics of Romanticism,* p. 78; and Lesley Sharpe, *Friedrich Schiller: Drama, Thought, and Politics* (Cambridge: Cambridge University Press, 1991), 161.

60. Michael Goldman, *Shakespeare and the Energies of Drama* (Princeton: Princeton University Press, 1972), p. 4.

61. Jean-Paul Davoine, "L'Epithète mélodramatique," *Revue des Science Humaines,* 162, no. 2 (1976), 183–186.

62. For the conjunction of sensuality with the "slovenly" look of paintings in the language of criticism, see Sam Smiles, "'Splashers,' 'Scrawlers,' and 'Plasterers': British Landscape Painting and the Language of Criticism," *Turner Studies,* 10 (Summer 1990), 5–11. On Danby, see Francis Greenacre, *Francis Danby (1793–1861),* exh. cat. (Bristol: Museum and Art Gallery; London: Tate Gallery, 1988), p. 101.

63. Nina Athanassoglu-Kallmayer, *Eugène Delacroix: Prints, Politics, and Satire, 1814–1822* (New Haven: Yale University Press, 1992), p. 108.

64. On superstition as an anthropological discourse, see Tobin Siebers, *The Romantic Fantastic* (Ithaca and London: Cornell University Press, 1984), pp. 12, 21, 35.

65. Ibid., p. 111. Also see Jerome Christensen, *Lord Byron's Strength: Romantic Writing and Commercial Society* (Baltimore and London: Johns Hopkins University Press, 1993), pp. 277, 284–291.

66. Michele Hannoosh, *Baudelaire and Caricature: From the Comic to an Art of Modernity* (University Park: Pennsylvania State University Press, 1992), pp. 4–6. On Delacroix's Stendhalian attempts to temporalize the ideal, see James H. Rubin, "Delacroix's *Dante and Virgil* as a Romantic Manifesto," *Art Journal,* 52 (Summer 1993), 52.

67. His work resembles Benjamin Constant's *The Spirit of Conquest and Usurpation and Their Relation to European Civilization* (1813). For Constant, see Jerome Christensen, "Byron's *Sardanapalus* and the Triumph of Liberalism," *Studies in Romanticism,* 31 (Fall 1992), 335–336.

68. Pascal Griener, "L'Art de persuader par l'image sous le 1er empire. A propos d'un concours officiel pour la représentation de Napoléon sur le champ de bataille d'Eylau," *L'Ecrivoir* (1984), 4, 9–11.

69. Robertson, *Mémoires récréatifs,* I, 145–149; and Salverte, *Des sciences occultes,* II, 256.

70. Michael Paul Driskel, "Manet, Naturalism, and the Politics of Christian Art," *Arts Magazine,* 60 (November 1985), 44–54; and *Representing Belief: Religion, Art, and Society in Nineteenth-Century France* (University Park and London: Pennsylvania State University Press, 1992), pp. 43, 49. Also see Gabriel P. Weisberg, "From the Real to the Unreal: Religious Painting and Photography at the Salons of the Third Republic," *Arts Magazine,* 60 (December 1985), 58–63; and Said, *Orientalism,* pp. 138–140.

71. Gossman, *Between History and Literature,* p. 194.

72. Herding, *Courbet,* pp. 157–160.

73. Mario Portigliatti Barbos, "Cesare Lombrosos delinquenter Mensch," in *Wunderblock,* pp. 587–588.

74. On Bourgery, see Reinhard Hilde-
brand, "Anatomie und Revolution des
Menschenbildes," *Sudhoffs Archiv,* 76,
no. 1 (1992), 1–6.

75. Michael Fried, *Courbet's Realism*
(Chicago: University of Chicago Press,
1990), p. 241.

76. Hazlitt, "The Indian Jugglers,"
pp. 219–223.

77. Norman Fruman, *Coleridge: The
Damaged Archangel* (New York:
George Braziller, 1971), pp. 39, 59,
69–70.

78. Saisselin, *Enlightenment against Baroque,* p. 64.

79. McGann, *Romantic Ideology,*
pp. 38–41.

80. Donald Kuspit, "The Problem of
Art in the Age of Glamour," *Art Criticism,* 6, no. 1 (1989), 33–36.